AF255794

THE WEAPONIZATION OF MIND CONTROL

THE WEAPONIZATION OF MIND CONTROL

A CAUTIONARY JOURNEY FROM ANCIENT RITUALS TO AI-DRIVEN MANIPULATION

JOHN SYDNEY SMITH

First Edition

ISBN: 978-1-7644355-0-5

Cover design and layout by John Sydney Smith

Independently published by John Sydney Smith

[www.johnsydneysmith.com

For permissions or more information, contact:

johnssmith@outlook.com.au

CONTENTS

PREFACE

From the dawn of history, human beings have sought ways to influence and control the thoughts, beliefs, and behaviors of others. This book explores that long, complex journey—from ancient rituals and primitive surgeries to modern techniques involving chemical agents, brain stimulation, and, ultimately, artificial intelligence. In recounting this evolution, I aim not only to document how mind control has been weaponized for both noble and nefarious ends but also to challenge us to consider the ethical implications of emerging neurotechnologies and AI.

Over millennia, societies have embraced practices intended either to uplift or to dominate. Rituals invoking divine inspiration have inspired communal values, while state-sponsored programs—often shrouded in secrecy—have exploited the vulnerabilities of human psychology. The atrocities committed in places such as Japan's Unit 731, where victims were reduced to mere data points in horrific experiments, force us to ask: How can ordinary people become instruments of collective evil? And what conditions allow authority, dehumanization, and propaganda to override our innate moral judgments?

This inquiry is more than historical; it resonates with our current age. Today, digital platforms, algorithmic decision-making, and advanced brain–computer interfaces are creating new arenas in which the human mind can be manipulated. As we navigate a future where our thoughts might be shaped not only by our immediate environment but also by engineered systems, understanding the past becomes essential. An overview of the milestones in the evolution of mind control is presented in Appendix I.

In the pages that follow, you will read about the methods once used to alter minds—the psychological manipulations, the chemical agents, the surgical procedures—and then journey forward into a discussion of cognitive warfare in the digital era. By drawing connections between historical abuses of mind control and today's challenges, I hope to provoke a thoughtful dialogue on how we might protect our most intimate faculties: our thoughts, emotions, and ultimately, our humanity.

This book is both a cautionary tale and a call to action. The legacy of mind control teaches us that power over the mind is perhaps the most profound power of all. As technology accelerates, and the boundaries between biological and digital intelligence blur, we must ask: Who controls the controllers? And how do we ensure that the tools designed to enhance our lives do not become instruments of oppression?

May this work serve as a foundation for understanding—and a warning about—the paths that have been trodden and those that lie ahead.

INTRODUCTION - THE ANATOMY OF A MASSACRE

On the morning of Saturday, March 16, 1968, the day began like any other in the village of Sơn Mỹ, South Vietnam. In the hamlet the Americans later called "My Lai 4," old men rose to boil rice and tend the small fires, women prepared baskets for the market, and children trailed behind mothers or played in the dust beside the family pigs. The war was never far away, but it was something that moved at the edges of their lives: distant artillery at night, passing patrols, the occasional helicopter thudding across the sky.

Just after dawn, that thudding sound came closer.

The first sign that this morning would be different was the shudder of artillery and the roar of helicopter rotors coming in low over the paddies. People stepped out of doorways to look up. Some assumed it was another search operation like others they had seen, when soldiers had passed through, taken a few men for questioning, handed sweets to the children, and moved on.

Instead, gunships raked the fields and hedgerows with machine-gun fire to "soften" the landing zone, even though there was no incoming fire from the ground. Villagers scattered in panic. Some ran for their family shelters—deep earthen bunkers dug beside their homes to protect against bombs. Others tried to flee across the open paddies, bent low, herding children ahead of them.

One boy later remembered his mother yanking him and his siblings toward the shelter as three American soldiers approached the house. One soldier set fire to the thatched roof. The others pulled the family into the bunker and hurled grenades in after them. When he regained consciousness hours later, he was buried under the bodies of his mother and four siblings, covered in their blood and hair. He was ten years old.

Everywhere in the hamlet, similar scenes unfolded. Soldiers from Charlie Company fanned out through the lanes, kicking in doors, dragging people from houses, shooting livestock, and setting huts alight. There were no Viet Cong in sight. The only people they encountered were old men, women, and children.

At first, some villagers believed they were being gathered for questioning or relocation. Groups were herded together in courtyards and along dikes, watched over by young Americans in dirty fatigues and steel helmets. But the orders those men were hearing were very different: they had been told that everyone remaining in the village was Viet Cong or Viet Cong sympathizers, and that they were to "kill anything that moves."

In one lane, an elderly man emerged from a hiding place with his hands raised, trying to surrender. A soldier drove a bayonet into his body. In another yard, villagers knelt in front of a small temple, burning incense and praying aloud; they were shot in the head where they knelt. Elsewhere, soldiers forced families into a bomb shelter, tossed grenades inside, and then fired into the opening.

Near the center of My Lai 4, dozens of villagers—mostly women and children—were marched to the edge of an irrigation ditch. They were ordered down into the trench and made to crouch or sit, shoulder to shoulder, in the mud. Some tried to shield their children with their bodies.

Then the shooting began.

Under orders from Lieutenant William Calley, American soldiers fired volley after volley into the crowded ditch, the crack of M16s rolling up and down the line. People screamed, prayed, scrambled over one another; then gradually the cries faded, replaced by the moans of the wounded and the thin wailing of trapped children. Over the next hour, more groups were driven to the same ditch and gunned down, until the trench was a long, reeking mass of bodies, five and six deep for more than thirty yards.

One witness later recalled seeing a small child, about three years old, claw his way up from under his dead mother, clamber out of the ditch, and stagger into the paddy. A soldier shouted that a child was escaping. When one of Calley's men hesitated to fire, Calley himself grabbed the child, threw him back into the ditch, and shot him.

Elsewhere in the hamlet, soldiers were bayoneting infants, shooting people at point-blank range in their homes. At least twenty women and girls—some as young as ten—were gang-raped, sodomized, or sexually tortured before being killed, their bodies later found with clothing ripped away and genitals mutilated. Some soldiers carved and hacked at corpses, cutting off ears or hands, slitting bellies, or scalping the dead.

Alongside the shooting and bayoneting, some soldiers descended into an almost animalistic frenzy. Witnesses later described one man repeatedly bayoneting a small boy, tossing him into the air and skewering him again "as if he were a papier-mâché piñata"–a grotesque game played with a living child, emblematic of the orgy of violence that engulfed parts of Mỹ Lai. Such behaviour cannot be explained by obedience alone.

It is also likely that some of the men were under the influence of mind-altering drugs. By 1971, a Department of Defense survey found that roughly half of US troops in Vietnam had used marijuana, about a third had tried psychedelics such as LSD, mescaline, or psilocybin, and more than a quarter had taken harder drugs like cocaine or heroin. Congressional investigations reported the military's consumption of hundreds of millions of stimulant tablets in a few years, chiefly amphetamines issued to improve endurance on patrol. Veterans later recalled that these pills not only increased alertness but also heightened irritability and aggression; some said that when the "speed" wore off they felt angry enough to shoot "children in the streets." In that context, it is not difficult to imagine that, at Mỹ Lai, fatigue, fear, and drug-induced disinhibition combined to lower self-control against extreme cruelty.

In the same fields and lanes, other Americans behaved very differently.

Some soldiers fired into the ditch because they believed refusal of an order under fire was punishable by court martial or even summary execution. Some did the minimum and then drifted away, smashing pots, killing animals, or simply wandering in a daze among the burning huts. A few refused outright: one machine gunner, ordered to shoot a group of villagers, lowered his weapon and said he would not; another, rather than continue in the killing, deliberately shot himself in the foot so he would be evacuated from the operation.

From above, the scene looked different again.

Warrant Officer Hugh Thompson, a 25-year-old helicopter pilot, was flying a reconnaissance mission over the area with his two-man crew when he began to grasp what was happening below. From the air he could see bodies scattered in the paddies and, later, the long trench choked with dead and dying villagers. At first, he assumed this was the aftermath of a firefight. But when he saw American soldiers firing into groups of unarmed civilians, including women and children, and shooting wounded people who were trying to crawl away, it became clear this was something else.

Thompson set his helicopter down near one group of survivors and confronted Calley on the ground, asking what was going on and offering to help evacuate the wounded. Calley brushed him off; later, Thompson learned that the wounded he had marked for evacuation had been executed.

A short time later, Thompson and his crew spotted another group of terrified villagers—about fifteen or sixteen people, including children—fleeing across a field, with American soldiers moving toward them. Thompson made a decision that would define his life: he flew his helicopter down and landed it between the soldiers and the civilians, using the aircraft as a physical barrier. He ordered his door gunner, Lawrence Colburn, and crew chief, Glenn Andreotta, to train their machine guns on their fellow Americans and to open fire if any soldier tried to harm the villagers. Then he coaxed the Vietnamese out and arranged for additional helicopters to airlift them to safety.

Circling back over the ditch, Thompson again saw movement among the pile of bodies. He landed once more. Andreotta climbed down into the bloody trench and emerged carrying a small, dazed girl, smeared with blood but still alive. They flew her out to medical care before returning to base, where Thompson reported to his superiors that a massacre of civilians was under way.

By late morning, the killing had largely stopped. In a few hours, more than 500 unarmed villagers lay dead in and around Mỹ Lai and a neighboring hamlet—about a third of them children. Among the 125 or so men of Charlie Company, there had been a spectrum of responses: a small core who seemed to take pleasure in the killing; a larger group who shot because they were ordered to; some who drifted at the edges or removed themselves; and a tiny minority who, like Thompson and his crew, risked their lives to protect the Vietnamese.[1]

It is in that spectrum—from cold enthusiasm through fearful obedience and passive complicity to active moral courage, all under the shadow of fatigue, fear and mind-altering drugs—that the anatomy of a massacre, and the possibilities of mind control and resistance, can be traced.

PART 1 - THE PSYCHOLOGICAL ASSAULT ON THE MIND

"The most potent weapon in the hands of the oppressor is the mind of the oppressed." – Steve Biko

CHAPTER 1- UNDERSTANDING EVIL

Introduction

Humankind's capacity for evil has perplexed philosophers, historians, and scientists for centuries. While individual antisocial acts—such as theft, assault, or murder—may stem from a combination of genetic predisposition and environmental hardship, explaining how ordinary people become complicit in collective atrocities presents a far more complex challenge. Can everyday individuals, given the right circumstances, commit unspeakable acts? Or do specific social, political, and cultural forces override personal morality? This chapter will explore these questions by examining historical case studies and the psychological mechanisms that enable group evil, laying the groundwork for later discussions of modern mind control and digital manipulation.

From Individual Acts to Collective Atrocity

At its most basic level, evil can be defined as actions that grossly violate accepted moral norms and inflict harm on others. For an individual, such behavior might be explained by a volatile combination of personal history, mental distress, and biological factors. However, history demonstrates that large-scale atrocities, whether perpetrated by groups, nations, or organized regimes, require more than individual pathology. They demand a context where dehumanization, obedience to authority, and groupthink converge to override the ethical intuitions of many.

Unit 731: a Case Study of Group Dehumanization

One of the most notorious examples of state-sponsored atrocities is the operation of Unit 731 during the Sino-Japanese War and World War II.[2] [3] Conceived by bacteriologist Professor Shirō Ishii, Unit 731 functioned under a regime that showed a profound disregard for human life. It became a massive apparatus of inhumane experimentation, where prisoners—including both enemy soldiers and civilians—were systematically dehumanized and subjected to lethal experiments. These included: deliberate infection with deadly pathogens using contaminated fleas, food and water to test biological weapons; infusions with animal blood and other substances in the search for a blood transfusion substitute; tethering to posts in an arna into which biological bombs were dropped to determine the most efficient design of such bombs, and horrific surgical procedures performed without anesthesia. Ultimately each victim underwent non-consensual human vivisection—live dissection of conscious prisoners without anaesthesia.

The purported aim of these atrocious experiments was two-fold: to find ways of protecting the Japanese military from wounds, blood loss, frostbite and plagues; and to create biological weapons of mass destruction. These weapons were designed not merely to kill individual enemies, but to unleash large-scale epidemics, contaminate water and food supplies, and render whole regions uninhabitable for prolonged periods. In practice, their deployment caused indiscriminate suffering: devastating outbreaks of plague, cholera and other diseases among civilian populations and troops alike, leading to mass death, long-term disability, psychological terror, and the collapse of already fragile public health systems.

The horrors of Unit 731 were not the actions of a few aberrant individuals acting independently but the result of a highly organized system that normalized cruelty and fostered obedience to authority. Researchers, military officers, and support staff collectively participated in macabre experiments characterized by dehumanization. The victims were referred to by numbers or as "Maruta" (meaning "logs") as the facility was previously a lumber mill; dehumanizing humor was encouraged by exchanges such as "How many logs did you down today?" This case vividly illustrates how collective evil emerges when individuals are subsumed into a group that shifts moral responsibility away from the self.

Unit 731: Historical Atrocities and Their Legacy

Although exact figures on the number of victims killed in Unit 731 remain unknown, estimates range from 3,000 to 12,000. There were no documented survivors, and the average life expectancy of prisoners upon entering Unit 731 was just two months. The main facility, located in Pingfang, Manchuria, was staffed by experts recruited from universities and research institutions. The initial group included eight assistant professors and instructors from Kyoto Imperial University.

The complicity of Unit 731 staff is exemplified in the testimony of a former member who participated in vivisections. As reported by Amelia Hill in a 2003 article for *The Guardian*[4]:

> The 'logs' we used were all very quiet because they had been infected with the germs and monitored until they were nearly dead. When you started to cut them, they usually died straight away …The first time I didn't know

what to do. I knew I had to kill it, but my body couldn't do it. Maybe from the third person, I started understanding what was going on and why it had to happen, and my legs didn't shake anymore. I remember the leader of the group told me I was getting better. Then I was no longer a child. I had finally become useful and efficient. I think I felt proud when he told me that.

Staff behavior revealed complexities—while many appeared to suspend moral convictions, they may have been motivated by ambitions in wartime research or career advancement. After the war, numerous Unit 731 personnel went on to assume prominent roles, including:

- Shirō Ishii was the mastermind behind the experiments. Postwar reports and secondary sources state that Ishii cooperated with US authorities and advised at Fort Detrick, Maryland, US, on Japanese biological warfare findings. [5]

- Prince Tsuneyoshi Takeda has been reported as having oversight responsibilities within the army command structure over Unit 731. [6] He was President of the Japanese Olympic Committee (1962–1969) and a member of the International Olympic Committee from 1967 to 1981. [7]

- Masaji Kitano (second-in-command of Unit 731), Ryoichi Naito (head of the germ warfare division), and Hideo Futaki (vivisection team leader) together co-founded Green Cross, Japan's first commercial blood bank.

- Hisato Yoshimura oversaw frostbite research, freezing the limbs of his victims—including babies—then testing w of thawing them out.[8] He became a physiology professor and in 1978 received the Order of the Rising Sun, Third Class. from Emperor Hirohito for his work on "environmental adaptation science."[9]

Psychological Mechanisms Underpinning Group Evil

From Individual to Collective Atrocity

The transformation from individual morality to collective atrocities is driven by well-documented psychological phenomena. Stanley Milgram's studies demonstrated that individuals, under conditions of high authority, could commit acts they would otherwise find repugnant. In Unit 731's case, systematic dehumanization of victims and pressure to conform to state ideology fostered a "moral vacuum," allowing participants to suspend personal ethics. This process—often described as "othering"—enabled participants to perceive victims not as humans but as obstacles to national or ideological objectives.

Lessons for Modern Society

Unit 731 represents a dark chapter in history, but its lessons are relevant to contemporary challenges. The psychological vulnerabilities that enabled such atrocities can be exploited today through advanced technologies, including digital manipulation, propaganda techniques, neurotechnologies, and artificial intelligence. Understanding the roots of collective evil is vital for building societal resilience and safeguarding human dignity in the face of modern threats.

Conclusion

Exploring the nature of evil—from its individual origins to collective manifestations—not only provides insight into history but also offers a framework for addressing potential future abuses of mind control. The case of Unit 731 and its psychological underpinnings set the stage for subsequent chapters, where we will trace the evolution of mind control techniques and confront the ethical challenges posed by emerging technologies. Recognizing these patterns is essential to ensure past mistakes are not repeated.

CHAPTER 2 - GROUP DYNAMICS

Introduction

How does a person's sense of right and wrong change when they join a group? History demonstrates that, under specific conditions such as unquestioned authority and routine "othering," ordinary individuals can contribute to atrocities. This chapter explores the psychological mechanisms underlying group dynamics.

The Psychology of Group Behavior

Group dynamics describe how people's thoughts, behaviors, and emotions shift when acting as part of a group. Early research in social psychology revealed that group settings often lead to phenomena such as conformity, obedience, and deindividuation. Under these conditions responsibility becomes diffused, so that individuals may view themselves as mere cogs in a larger machine, feeling less accountable for actions that violate their ethical standards.

Diffusion of Responsibility

A key concept in group dynamics is the "diffusion of responsibility." In group settings, decisions and actions are distributed among many members. Consequently, even when harmful behavior occurs, individuals may not feel personally responsible, assuming others will share the blame. This dynamic contributes to increased group violence and moral violations.

Landmark Studies in Group Dynamic

Sherif's robbers cave experiment

Muzafer Sherif's 1954 Robbers Cave Experiment remains one of the most influential studies in this field.[10] Boys at a summer camp were divided into two groups and placed in a competitive environment. Initially, each group—unaware of the other's existence—developed its own social norms and hierarchy. When introduced to the other group, competition resulted in hostility. However, cooperation toward shared "superordinate goals" ultimately reduced hostilities. This study illustrates how group identity can foster both conflict and reconciliation, depending on whether the focus is on division or unity.[11]

Milgram's Obedience Studies

Stanley Milgram's groundbreaking experiments on obedience revealed how social pressure shapes behavior. Participants were willing to administer what they were told were painful electric shocks to others simply because an authority figure insisted. Milgram's findings suggest that the power of authority can override personal morals, especially when individuals feel they are merely following orders.[12]

Solomon Asch's Conformity Experiments

In the 1950s, psychologist Solomon Asch conducted a series of classic experiments to test how strongly group pressure can influence individual judgment. In his basic setup, a small group of people were shown a "target" line and then three comparison lines, one of which clearly matched the target in length. All but one of the people in the room were confederates instructed to give obviously wrong answers on most trials, while a single naive participant believed everyone was taking part in a simple vision test.

When participants made their judgments alone, they chose the correct line almost every time. In the group setting, however, Asch found that roughly three-quarters of participants conformed to the group's wrong answer at least once, and on average they went along with the incorrect majority about one-third of the time across the "critical" trials. Afterward, many explained that they had doubted their own perception or gone along simply to avoid standing out. Asch also showed that even one ally in the room who gave the correct answer dramatically reduced conformity, suggesting that social support can strengthen people's willingness to resist group pressure.

Asch concluded that the desire to fit in and the tendency to assume that "the group must be right" can lead people to publicly endorse claims they know are false. His findings remain a foundational demonstration of how powerful conformity pressures can become—especially in cohesive groups where dissent is discouraged and leaders signal clear expectations about what it means to be "on the team."

Zimbardo's Stanford Prison Experiment

Philip Zimbardo's Stanford Prison Experiment purportedly demonstrated how situational factors, rather than personality traits, drive group behavior. [13] Participants randomly assigned roles as guards or prisoners began exhibiting cruel or submissive behaviors consistent with their roles. However, later investigations revealed that participants knew what Zimbardo was trying to prove and acted accordingly. Psychologists Steve Reicher and Alexander Haslam failed to replicate these findings, concluding instead that tyranny and cruelty depend on active, ideologically motivated leadership plus group processes, not on people "slipping mindlessly into roles."[14]

Deindividuation

In his 1895 work *The Crowd: A Study of the Popular Mind,* polymath Gustave Le Bon proposed that when they are in a crowd individuals behave differently, because crowds create a "collective mind" set that allows them to act emotionally, irrationally, and often contrary to their usual behaviour.[15] He proposed that there were three elements that determined this, firstly anonymity which promoted a loss of personal responsibility, secondly contagion so that this loss spread across the group, and thirdly the fact that people in crowds are more suggestible.[16]

Le Bon's writings led to the study of deindividuation, defined as the psychological state in which individuals lose their sense of self-awareness and personal identity when in a group, and indulge in uncharacteristic behaviour that can be altruistic (e.g., protesting for social reform) or antisocial (such as rioting or vandalism).

In contemporary times, online platforms that allow anonymity or fake accounts can foster inflammatory or offensive comments that may be out of character.[17] In military or wartime settings, soldiers in uniform often comply with unit norms and orders, committing acts they would not choose as isolated individuals.[18]

Mechanisms of "Othering" and Moral Transformation

The Process of "othering"

A recurring theme in these studies is "othering"—defining an in-group in opposition to an "out-group." By categorizing others as different or inferior, group members justify actions otherwise deemed immoral. Historical events, from colonization, to wartime atrocities to political purges, demonstrate how "othering" often serves as a foundation for collective violence.

Shifts in Moral Responsibility

In group contexts, individuals may suspend personal moral judgments. Moral decisions made collectively or dictated by authority figures often compromise individual ethical standards. High-pressure environments blur traditional boundaries between right and wrong, creating conditions for actions that, in isolation, would seem unthinkable.

Modern Implications and Ethical Reflections

The insights gleaned from these experiments extend beyond academic inquiry; they offer valuable lessons for understanding modern forms of manipulation. Today, digital platforms and algorithm-driven media frequently recreate conditions like those observed in the laboratory. For example, online echo chambers—

which expose individuals to information that reinforces preexisting beliefs—and targeted propaganda exploit the same psychological mechanisms of conformity and diffusion of responsibility to shape public opinion. As technology evolves, so too does its potential for large-scale cognitive manipulation.

These developments raise urgent ethical questions. How can individual autonomy be protected in environments designed to suppress dissent or promote specific narratives? What responsibilities do leaders—whether in politics, business, or technology—bear in mitigating the risks inherent in group dynamics?

Conclusion

Group dynamics reveal a fundamental truth: the context in which we operate dramatically influences our perceptions and actions. The research of Sherif, Milgram, Reicher, and Haslam demonstrates that under the right conditions, even well-meaning individuals can engage in harmful behavior. Recognizing these vulnerabilities is vital, not only for understanding historical atrocities but also for addressing contemporary challenges like digital propaganda and cognitive warfare.

As we delve deeper into mind control techniques, the lessons of group dynamics emphasize that preserving individual moral judgment requires constant vigilance. Future chapters will explore how these psychological forces have been weaponized—through propaganda and other methods—and examine the modern technological threats they pose.

CHAPTER 3 - PROPAGANDA

Introduction

Propaganda is not a new phenomenon. For centuries, rulers, armies, and ideologues have harnessed information—and misinformation—to shape public opinion, justify policies, and mobilize entire populations. This chapter traces the evolution of propaganda from its classical origins through its development during modern warfare, and finally to its sophisticated forms in the digital age. By understanding the methods and effects of propaganda, we can better grasp its potential for both mass mobilization and mass manipulation.

The Historical Evolution of Propaganda

Early Beginnings

Ancient leaders such as Augustus and Napoleon recognized that controlling the narrative was as crucial as controlling territory. Augustus (63 BCE - 14 CE), for instance, distributed inscriptions listing his accomplishments throughout the Roman Empire to cultivate a benevolent image. Similarly, Napoleon (1769 – 1821) commissioned art and monuments celebrating his military victories, effectively rewriting history in his favor. These early examples highlight the core function of propaganda: to shape perception by selectively presenting information.

Propaganda in Modern Warfare

The devastation of World War I marked a dramatic leap forward in propaganda techniques. Governments mobilized entire populations using a mix of art, film, and printed media to craft persuasive narratives that often blurred the lines between fact and fiction. Posters urged citizens to enlist or make sacrifices, while state-controlled news outlets amplified messages designed to demonize the enemy. [19] The intensity and coordination of these campaigns laid the groundwork for propaganda's role in totalitarian regimes, where it later contributed to the justification of atrocities.

Core Techniques of Propaganda

Repetition and Simplification

As Adolf Hitler theorized in his 1925 manuscript *Mein Kampf* (*My Struggle*), and as Joseph Goebbels demonstrated in developing the Third Reich, effective propaganda relies on repetition and the simplification of complex issues into clear, memorable slogans. This approach exploits the human tendency to internalize repeated information, even when it is incomplete or misleading. When a message is repeated often enough, it begins to take on a veneer of truth, regardless of its factual accuracy.

In the Third Reich, an estimated 125,000 posters were distributed weekly, displayed in public spaces and transport venues. These posters extolled the virtues of the state and the heroism of Nazi youth while decrying perceived societal evils.

Concepts were reduced to simple labels and slogans: the Führer was portrayed as a god-like figure, Churchill and Roosevelt as "warmongers," Aryans as the "Master Race," and Jews as "The Jewish Parasite."[20] Caricatures and captions such as "Who is to blame for the War?" reinforced these narratives. The eugenics program was similarly promoted through posters depicting the infirm with captions like, "This genetically ill person will cost our people's community 60,000 marks over his lifetime."

Symbols played a significant role. While writing *Mein Kampf*, Hitler designed the Nazi flag, featuring a red background with a white disc and a black swastika. Swastika banners became omnipresent, and uniforms were adorned with swastika armbands.

Hitler's speeches were broadcast through millions of radios sold cheaply via government subsidies and through loudspeakers in public spaces and workplaces.[21] Books and magazines were censored, while those promoting German nationalism, anti-Semitism, or eugenics were widely distributed. All newspapers were placed under Nazi control and were required to promote Nazi ideology. To discredit critics, the Nazis used the slogan "Lügenpresse" ("lying press"), a term echoed in modern rhetoric like "fake news."

Emotional Appeals

Propaganda frequently leverages strong emotional triggers—fear, pride, anger, or hope—to bypass critical thinking. By tapping into deep emotions, propagandists can override rational evaluation. This emotional manipulation is evident in wartime posters evoking patriotism and fear of the enemy, as well as in modern campaigns using imagery and music to elicit immediate, visceral responses.

In the lead-up to World War II, the annual Nuremberg Party Rallies galvanized the population. Tens of thousands gathered in distinctive uniforms, encircled by the 'Cathedral of Light' formed by more than 130 anti-aircraft searchlights whose beams reached miles into the night sky,"[22] The rallies opened with Wagner's *Rienzi* overture, and choirs sang songs glorifying the Reich. On occasion, Hitler descended *deus ex machina*, landing his plane in the arena.

Hitler's speeches were practiced to perfection. A prolonged silence, designed to create an air of anticipation, was followed by a softly spoken introduction with his body still, which gradually built to a crescendo with raised voice, contorted face and wild but controlled gestures as he railed against the iniquities suffered, especially at the hands of the Jews. Finally, suddenly, he would fall silent, leaving his audience in rapture.

The Nazis also used film to manipulate emotions. Leni Riefenstahl's 1933 film *Nuremberg Rally* and her 1934 film *Triumph of the Will* glorified the Reich, while Fritz Hippler's 1940 film *The Eternal Jew* portrayed Jews as unclean, soulless, and exploitative criminals.

Dehumanization and "othering"

As the Nazis demonstrated, one of the most potent strategies in propaganda is dehumanization—the process of casting a group as "the Other." By stripping away individuality and presenting a target group as homogeneous and morally inferior, propagandists make it easier to justify extreme measures against them.[23] This tactic has been central to some of history's most devastating conflicts and continues to be used in modern political rhetoric.

Propaganda in the Digital Age

Today's propaganda has evolved far beyond posters and radio broadcasts. Digital platforms now enable the rapid dissemination of information—and disinformation—on a global scale. Social media algorithms, designed to maximize engagement, can inadvertently or deliberately reinforce echo chambers. Users are often exposed to repetitive, emotionally charged content that mirrors historical propaganda techniques. For instance, targeted advertising and deepfake videos are increasingly used to subtly and pervasively manipulate public perception.[24]

Ethical Implications

The evolution of propaganda raises profound ethical concerns. With digital tools, the capacity to influence public opinion has become both more powerful and more insidious. When artificial intelligence algorithms curate and amplify messages without transparency, they exacerbate social divisions and erode democratic processes. The ethical imperative to ensure that technology serves the public good—rather than being used as a tool for manipulation—has never been more urgent.

Conclusion

Propaganda, in all its forms, is a double-edged sword. Its ability to unite or divide, to inspire or incite, has been demonstrated time and again throughout history. From the carefully crafted messages of ancient emperors to the algorithm-driven narratives of the digital era, propaganda remains a key tool for shaping human behavior. Understanding its evolution, techniques, and ethical implications is essential—not just for interpreting the past but also for safeguarding the future.

The chapters that follow will explore how these principles have been applied in more direct methods of mind control—from chemical and surgical interventions to modern digital manipulation. Recognizing the persistent influence of propaganda is the first step in building resilience against its misuse.

Adolf Hitler's Propaganda Manual

"If a movement proposes to overthrow a certain order of things and construct a new one in its place, then the following principles must be clearly understood and must dominate in the ranks of its leadership: Every movement which has gained its human material must first divide this material into two groups: namely, followers and members ... To be a follower needs only the passive recognition of the idea. To be a member means to represent that idea and fight for it ... Because of its passive character, the simple effort of believing in a political doctrine is enough for the majority, for the majority of mankind is mentally lazy and timid. To be a member one must be intellectually active, and therefore this applies only to the minority ...

The purpose of propaganda is not the personal instruction of the individual, but rather to attract public attention to certain things, the importance of which can be brought home to the masses only by this means ... All propaganda must be presented in a popular form and must fix its intellectual level so as not to be above the heads of the least intellectual of those to whom it is directed ...

Once we have understood how necessary it is to concentrate the persuasive forces of propaganda on the broad masses of the people, the following lessons result therefrom: That it is a mistake to organize the direct propaganda as if it were a manifold system of scientific instruction. The receptive powers of the masses are very restricted, and their understanding is feeble. On the other hand, they quickly forget. Such being the case, all effective propaganda must be confined to a few bare essentials and those must be expressed as far as possible in stereotyped formulas. These slogans should be persistently repeated until the very last individual has come to grasp the idea that has been put forward ...

Propaganda must not investigate the truth objectively and, in so far as it is favourable to the other side, present it according to the theoretical rules of justice; yet it must present only that aspect of the truth which is favourable to its own side ... Propaganda must be limited to a few simple themes and these must be represented again and again."

Mein Kampf 1926

CHAPTER 4 – SOCIETY'S OUTCASTS

Introduction

Throughout history, societies have defined themselves not only by their values but also by what they reject. Those who fail to conform to prevailing social, cultural, or political norms are often cast as "others"—outsiders whose differences justify exclusion, discrimination, or even violence. This chapter explores the mechanisms by which certain groups have been marginalized, examines the pseudoscientific and ideological tools used to dehumanize them, and reflects on how these practices persist in modern times.

The Process of "Othering"

At its core, "othering" is the systematic categorization of individuals or groups as fundamentally different from—and inferior to—the dominant group. By labeling certain traits as abnormal or dangerous, societies create a powerful "us versus them" dichotomy. This process, observed across cultures and eras, is central to justifying various forms of social control including colonization and genocide. "Othering" involves several interconnected psychological and social mechanisms:[25]

- Stereotyping and Simplification: Complex human differences are reduced to oversimplified traits. By distilling a diverse group into a single set of negative characteristics, the dominant group creates an image that is easier to vilify.

• Diffusion of Empathy: Viewing people as "others" makes it easier to suspend empathy. This lack of compassion paves the way for policies and actions that harm marginalized groups without moral restraint.

• Legitimization of Power: Designating certain groups as "outcasts" allows authorities to justify coercive measures—ranging from social exclusion to overt violence—under the guise of protecting the majority's security or well-being.

"Othering" and Eugenics

In 1883, British polymath Sir Francis Galton introduced the concept of "eugenics," asserting that intelligence was inherited, and that society should encourage the reproduction of "superior" individuals while discouraging those deemed less capable.[26] Derived from the Greek words "eu" (good) and "genēs" (born), eugenics quickly gained traction among civic leaders.

Prominent figures like Theodore Roosevelt[27] and Winston Churchill[28] endorsed eugenic ideas. Roosevelt warned of "race suicide" if eugenic measures were not implemented, while Churchill expressed concern over the "unnatural and increasingly rapid growth" of the "feeble-minded." In his 1916 book *The Passing of the Great Race*, US lawyer Madison Grant argued that immigration from "non-Nordic Europe" was contaminating the US gene pool. [29] Adolf Hitler later referred to Grant's work as "my bible" and it was cited at the Nuremberg Trials as evidence that eugenics ideas pre-dated Nazism and had deep roots in the US.

Eugenic policies often relied on pseudoscientific claims. While Charles Darwin did not support eugenics, proponents misused his theory of natural selection to argue that only the "fittest" should reproduce. [30] Public education campaigns, including pamphlets and lectures, spread these ideas widely.

Eugenics influenced immigration policies: Australia's Immigration Restriction Act of 1901, [31] barred non-white immigrants; the US Immigration Act of 1924 limited immigration from Southern and Eastern Europe. [32] The most insidious programs, however, involved the forced sterilization of those deemed "unfit."

In 1907, Indiana became the first US state to enact a eugenics law, mandating sterilization for individuals in state custody judged mentally unfit. [33] By the 1970s, when such laws were repealed, an estimated 60,000 Americans had been sterilized, with one-third of these cases occurring in California. [34] Similarly, Canada's Alberta Province passed the Sexual Sterilization Act in 1928, resulting in 2,832 sterilizations before its repeal in 1972. [35]

Germany's eugenics movement took a darker turn under the Weimar Republic and later the Nazi regime. The 1920 book *Die Freigabe der Vernichtung Lebensunwerten Lebens* (*Allowing the Destruction of Life Unworthy of Living*), by Karl Binding and Alfred Hoche, justified killing individuals deemed of no value to society. [36] The Nazis implemented compulsory sterilization laws targeting those with conditions such as schizophrenia, epilepsy, and severe alcoholism. Between 1933 and 1939, an estimated 360,000 to 375,000 Germans were forcibly sterilized. [37]

In 1939, Hitler authorized the systematic killing of psychiatric patients, resulting in the deaths of 200,000 to 300,000 individuals under the guise of "mercy killings."[38] This program expanded to include other marginalized groups, including those with disabilities, homosexuals, and ethnic minorities.

Eugenic sterilization laws were also enacted in countries such as France, Sweden, Denmark, and Brazil.[39] In Austria, sterilizations began only after the Nazi occupation, despite earlier advocacy by Nobel laureate Wagner-Jauregg.[40]

"Othering" and The Great Replacement Theory

"Othering" has played a significant role in shaping political and racial conflicts. According to recent United Nations and International Organization for Migration data, there were about 281–304 million international migrants worldwide (roughly 3.6–3.7% of the global population) in the early 2020s, and their number has nearly doubled since 1990, with most migrants originating from developing regions and continued growth expected in the coming decades.[41] While migration can benefit both host countries and migrants, it has historically provoked concerns about economic, cultural, and religious repercussions. Over the past century, these concerns have evolved into conspiratorial beliefs, suggesting that migration is being intentionally orchestrated to gain political advantage or to dilute cultural identity through interracial breeding.

Edward Alsworth Ross, an American sociologist and eugenicist, gained notoriety in 1900 for his opposition to Chinese and Japanese immigration. Coining the term "race suicide," he warned that the declining birthrate among "old-stock Americans" was leading to their extinction, outpaced by inferior but more fecund immigrant races.[42] His solution was to increase the fertility rate of white women—a proposal endorsed by President Theodore Roosevelt, who in 1905 described childless individuals as "criminals against the race" deserving of contempt.[43]

In the UK, figures like Sir Oswald Mosley capitalized on racial tensions to push extremist agendas. Mosley proposed halting immigration of people of color, repatriating Caribbean immigrants, and banning interracial marriages.[44] Similarly, Enoch Powell warned in 1968 that continued immigration would culminate in the "funeral pyre" of the British nation.[45]

France saw the emergence of replacement theory in Jean Raspail's 1973 novel *The Camp of the Saints*, a dystopian depiction of Western civilization's destruction through mass immigration from developing countries.[46] [47] The novel, which advocated violent resistance against non-white migrants, gained traction among far-right and white nationalist groups, influencing contemporary figures like France's far-right politician Marine Le Pen and Hungary's Prime Minister Viktor Orbán.

Echoes of Historical "Othering" in the 21st Century

The legacy of "othering" remains evident in contemporary society, amplified by digital platforms and algorithm-driven media that propagate divisive narratives. Social media echo chambers intensify fears of "us versus them," perpetuating alienation and distrust and often inciting real-world violence.

Renaud Camus, influenced by Raspail, published *The Great Replacement* in 2011, claiming that white Europeans were being "reverse colonized" by Black and Brown immigrants in an existential threat to Western society.[48] This ideology spread globally among alt-right and white supremacist groups, with its slogans — such as "You will not replace us" — becoming rallying cries at events like the 2017 "Unite the Right" rally in Charlottesville, Virginia, USA.[49]

The Great Replacement Theory gained mainstream attention in the US, with figures like journalist Tucker Carlson promoting it on Fox News. The theory claims that demographic shifts are deliberate strategies by liberal politicians to "replace conservative white voters." In a 2022 survey by the Southern Poverty Law Center, over a third of Americans expressed concern that these demographic changes threatened white culture and values.[50] Among Republicans, nearly seven in ten believed that such changes were politically motivated.[51]

Tragically, this ideology has inspired multiple acts of mass violence. In 2011, a Norwegian extremist killed 77 people, citing fears of Muslim population growth. Similar motives underpinned attacks like the Christchurch Mosque shooting in 2019, the El Paso Walmart massacre targeting Hispanics, and the Buffalo supermarket shooting in 2022.[52] [53] These incidents highlight the dangerous real-world consequences of the replacement theory.

Ethical Implications

The persistent use of "othering", particularly through the replacement theory, has profound ethical consequences. Policies rooted in exclusion or dehumanization violate fundamental rights and undermine societal cohesion. While modern applications may appear subtler than historical precedents, their impact remains equally harmful—enabling systemic discrimination and eroding democratic values.

The Role of Institutions in Mitigating "Othering"

Combating "othering" requires proactive efforts from educational, political, and social institutions. Recognizing its historical roots and psychological mechanisms is essential for fostering inclusivity. Institutions must prioritize transparency in policy, equitable resource distribution, and public dialogue that challenges stereotypes and promotes empathy.

Conclusion

From eugenics-driven sterilizations to totalitarian propaganda and migrant vilification, "othering" has been a tool of control and justification for atrocities. Understanding its historical construction—and how it persists in subtle forms—is vital for protecting human dignity. Vigilance against divisive forces and the promotion of social structures that emphasize mutual respect are essential in shaping a more inclusive future.

The lessons of history serve as both caution and inspiration. By rejecting narratives that categorize others as outsiders, we can work toward celebrating human diversity and harnessing the power of unity rather than division.

CHAPTER 5 – CHINA'S EXPERIMENT WITH BRAINWASHING

Introduction

In the mid-twentieth century, as the Chinese Communist Party (CCP) consolidated its power, it launched programs aimed at reshaping the thoughts, beliefs, and loyalties of its citizens. Often referred to as "brainwashing" or thought reform in Western scholarship, these efforts sought to eliminate dissent and enforce ideological conformity. This chapter explores the historical context, methods, and enduring impact of China's brainwashing experiments, offering ethical lessons that remain relevant today.

Historical Background

When the People's Republic of China was established in 1949, the CCP faced the daunting task of uniting a diverse population of around 540 million and rebuilding a nation devastated by war. To achieve political stability, the Party believed it should ensure that every citizen embraced its revolutionary ideals. Amid the Cold War tensions of the early 1950s, the CCP instituted state-sponsored "thought reform" programs designed to convert counterrevolutionaries, dissenters, and ordinary citizens into loyal Party supporters. [54]

China's early thought reform efforts combined intensive propaganda with psychological and physical coercion. Reeducation camps and "struggle sessions" became central to these campaigns.[55] Detainees endured prolonged isolation, public humiliation, and rigorous ideological instruction designed to dismantle preexisting beliefs and impose Party-approved narratives. Though many aspects of these programs remain shrouded in secrecy, survivor testimonies and declassified records provide insight into the systematic efforts to "reprogram" individuals in service of the state.

Techniques of Brainwashing

Isolation and Sensory Deprivation

Isolation was a key component of thought reform. In reeducation settings, detainees were separated from family, friends, and familiar surroundings, severing the social bonds that reinforced their individuality. Many subjects were held in overcrowded cells with minimal sensory input, rendering them more susceptible to repeated exposure to state propaganda.

Intensive Interrogation and Public Humiliation

Interrogation played a pivotal role in China's brainwashing efforts. Detainees were coerced into confessing crimes they had not committed. "Struggle sessions" further heightened the psychological toll, with subjects publicly humiliated and forced to admit their ideological "failings" under the scrutiny of Party officials. The process of public shaming and self-criticism eroded self-esteem and reinforced the idea that dissent was not only morally wrong but dangerous.

Propaganda and Repetition

Propaganda was deployed relentlessly to embed Party ideology within the population. Posters, radio broadcasts, and staged public events inundated individuals with simplistic slogans and emotionally charged narratives, making independent thought increasingly difficult. The saturation of state messages within a coercive environment often led many to internalize the new doctrine.

---

the legacy and modern parallels

While China's mid-century thought reform relied on physical isolation and face-to-face indoctrination, its principles resonate in contemporary contexts. Today, digital platforms and algorithm-driven media create virtual echo chambers, isolating individuals from dissenting viewpoints.

State surveillance systems, such as China's social credit system, extend the tools of control, enabling governments to monitor and manipulate behavior on a massive scale.

The legacy of thought reform—where the goal is ultimate control over thought—offers a sobering reminder of the dangers of unchecked power.

Lessons for Today

China's experiments with brainwashing reveal the susceptibility of the human mind to systematic coercion. In an era of rapid technological advancements, where artificial intelligence and neurotechnologies offer new avenues for influencing thought, the importance of safeguarding autonomy and promoting ethical transparency cannot be overstated. These historical abuses serve as urgent reminders of the need to protect individual freedoms against exploitation.

Conclusion

China's thought reform campaigns represent a dark chapter in the history of state control—one where isolation, coercion, and relentless propaganda were used to suppress dissent and create ideological uniformity. As we examine the evolution of mind control from ancient techniques to modern digital manipulation, the lessons of China's brainwashing experiments remain both cautionary and instructive. Recognizing the potential for propaganda to reshape perception and understanding the impact of psychological coercion are critical steps in safeguarding human dignity and freedom in an increasingly complex world.

The "Brainwashing" of Dr. Charles Vincent

Dr. Charles Vincent, a Western physician working in Shanghai, was one of 25 subjects interviewed by Robert Jay Lifton for his 1963 study *Thought Reform and the Psychology of Totalism: A Study of "Brainwashing" in China.* Arrested on allegations of espionage, Vincent was thrown into an eight-by-twelve-foot bare cell with eight other prisoners. His early detention was marked by relentless psychological and physical abuse—inmates surrounded him voicing accusations of imperialism and spying, while interrogators, including an official known only as the judge, repeatedly demanded that he confess his "crimes."

Over the next several days, Vincent was subjected to a brutal regimen: prolonged interrogations interspersed with physical punishments such as dragging him by his ankle chains and enforcing long periods of standing with his hands handcuffed behind his back. Sleep was denied for eight continuous days until, overwhelmed by fatigue and isolation, he produced a "wild confession" of fabricated espionage. This was followed by forced dictations of detailed accounts from his twenty years in China.

For three weeks, he was compelled to sign confessions with a thumbprint despite continued restraint. The process escalated into the recording of lists featuring everyone he had known, cementing the regime's aim to obliterate his personal identity.

After two weeks of further denunciation of his contacts, Vincent's ordeal shifted: he was permitted limited physical relief and, over 14 months, underwent a "re-education" program focused on condemning Western ideals and embracing Communist definitions of criminality. In his third year, with modest privileges such as outdoor exercise and roles teaching French and medicine, he was forced to participate in a public signing of his confession—an event widely publicized through photographs and film.

Upon his release, and subsequent expulsion from China, Vincent experienced an internal transformation—from initial tolerance and confusion to firm condemnation of his captors. Lifton later noted that although all the prisoners struggled to reclaim their identities after imprisonment, nearly all eventually rejected the imposed Communist ideology.

CHAPTER 6 - THE METAMORPHYSIS OF THE US REPUBLICAN PARTY

Introduction

The evolution of the US Republican Party over the past several decades is not merely a story of shifting policy positions—it is a case study in how political movements can be transformed through psychological techniques, propaganda, and the deliberate shaping of group identity. This chapter examines how the Republican Party, once rooted in conventional conservative principles, gradually adopted elements of radical rhetoric and mind-control strategies. In doing so, it not only redefined its own identity but also reshaped the nation's political landscape.

Historical Roots of a Changing Identity

Historically, the Republican Party championed fiscal responsibility, limited government, and a strong national defense. However, beginning in the mid-20th century, the party's base began to shift as new ideas emerged that prioritized emotional appeals over policy nuance. This transformation illustrates a profound change in how political groups mobilize support and shape ideology. By employing propaganda techniques, exploiting group dynamics, and fostering an environment of intense "othering", the Republican Party influenced the broader political climate of the nation.

From Traditional Conservatism to the New Right

The John Birch Society

In the early 20th century, wealthy families such as the Pews, Kochs, Olins, and Welsches played a pivotal role in shifting Republican ideology to the right. In December 1958, Henry Welch, Fred Koch, and their allies founded the John Birch Society (JBS), which claimed that America was in moral decline. The JBS opposed big government, the United Nations, and free trade agreements, while advocating for the dismantling of the Federal Reserve System. It condemned the Equal Rights Amendment, the civil rights movement, and feminism, while promoting conspiracy theories, including claims that President Dwight D. Eisenhower was a Communist and that world events were controlled by the Illuminati. [56]

The JBS disseminated its ideology through a Speaker's Bureau, magazines like *American Opinion* and *The New American*, newsletters such as the *JBS Bulletin*, and 2.5 million copies of Fred Koch's manifesto *A Businessman Looks at Communism*. [57]

By the mid-1960s, the JBS was spending millions of dollars annually to sponsor thousands of radio and television broadcasts promoting its anti-communist, anti-government message. Over subsequent decades, a broader conservative infrastructure expanded this media and educational influence: Koch-affiliated foundations directed hundreds of millions of dollars to programs on hundreds of university campuses,[58] while the John M. Olin Foundation disbursed more than $370 million over 25 years to law schools, think tanks, and media projects, subtly shaping curricula, staff appointments, and student activities in line with conservative ideals.[59]

Televangelism

By the 1970s, televangelism had become a powerful force in the US. Figures like Billy Graham, Pat Robertson, and Jerry Falwell preached a fundamentalist Christianity that emphasized biblical inerrancy and literal interpretations of scripture. The audience for syndicated religious television programs grew from under 10 million in 1970 to nearly 21 million by 1975.[60]

In 1979, Jerry Falwell and political activist Paul Weyrich co-founded the Moral Majority, aiming to "re-create this great nation" through political power.[61] [62] Evangelists directed their congregations on voting choices and encouraged grassroots mobilization. Ronald Reagan capitalized on this influence, appointing the Moral Majority's first executive director, Rev. Robert Billings, as a religious advisor to his campaign.[63]

The National Rifle Association

The National Rifle Association emerged as another potent political force. The Second Amendment, ratified in 1791, ambiguously addressed gun ownership within the context of a "well-regulated militia." Between 1876 and 1939, the US Supreme Court consistently ruled that the amendment did not protect individual gun ownership outside militia service. However, in 1977, the NRA reinterpreted the amendment to advocate for individual gun rights, funding legal opinions to support this view. In 2008, the Supreme Court ruled 5-4 that the Second Amendment guaranteed the right to own weapons "in common use" for home defense. [64]

Subsequently gun ownership became a political identifier. In a July 2024 Pew Research Center fact sheet, 45% of Republicans and Republican-leaning independents said they personally own a gun, compared with 20% of Democrats and Democratic leaners. It further noted that 58% of US adults favor making gun laws stricter than they are today, while 15% say laws should be less strict; 26% say laws are about right. [65]

The "New Right"

In the 1970s, a coalition of Moral Majority evangelists, the NRA, the anti-gay activist Anita Bryant, and the equal rights opposer Phyllis Schlafly coalesced into the "New Right." This movement, instrumental in Ronald Reagan's 1980 presidential victory, sought to advance conservative social and political goals.

The Federalist Society

Law students at Yale, Harvard, and the University of Chicago founded The Federalist Society in 1982 to promote conservative and libertarian legal ideas, with early support from conservative foundations linked to the Olin-backed Institute for Educational Affairs. The society has championed originalist and textualist approaches to the Constitution and supported conservative positions on abortion, LGBT rights, gun ownership, and federalism. By the late 2010s it counted more than 70,000 lawyers among its members and was widely regarded as the most influential conservative legal organization in the United States.[66] During Donald Trump's initial presidency, nearly 90% of his appellate court judges and both of his Supreme Court nominees were drawn from Federalist Society ranks, and by the early 2020s at least five—and arguably six—of the nine Supreme Court justices had current or former ties to the organization.[67]

The Tea Party

In 1984, Charles and David Koch founded Citizens for a Sound Economy, a conservative advocacy group promoting free markets, limited government, and grassroots involvement in public policy. In 2002, Citizens for a Sound Economy launched the website www.usteaparty.com to promote a continuous online "US Tea Party" for Americans who believed taxes were too high and the tax code too complicated.[68]

In January 2009, Barack Obama was inaugurated as the 44th president of the United States. The following month he announced the 75-billion-dollar Homeowner Affordability and Stability Plan, intended to help up to nine million homeowners avoid foreclosure during the financial crisis. The next day, CNBC reporter Rick Santelli, broadcasting from the Chicago Mercantile Exchange, denounced the plan for "subsidizing the losers' mortgages" and called for a "Chicago Tea Party" to protest government intervention. His rant went viral, and within hours the conservative group Americans for Prosperity registered

TaxDayTeaParty.com and launched a site calling for anti-tax, anti-spending protests. Within days, "tea party" events were organized in dozens of cities, leading to a coordinated nationwide protest on February 27. The modern Tea Party was born. By Tax Day that April, hundreds of demonstrations had taken place across the country. [69]

The Tea Party advocated for smaller government, reduced government spending, lower national debt, limited foreign aid, no tax increases, and American exceptionalism, framing its agenda as rooted in the Constitution. While most Tea Party supporters were Republicans or Republican-leaning independents, they frequently expressed frustration with mainstream Republican leaders and used primary challenges to back insurgent conservative candidates. Many of those who backed the Tea Party during the Obama years became some of Donald Trump's most enthusiastic supporters in 2016, and scholars argue that Tea Party activism and rhetoric helped pave the way for his rise and for the ongoing populist, hard-right turn of the Republican Party. [70]

Theocracy

Leonard Anthony Leo is regarded as one of the most influential figures in the contemporary Republican Party. After earning his Juris Doctor from Cornell Law School in 1989, Leo founded a student chapter of the Federalist Society during his final year. In 1991, at just 25 years old, he joined efforts to secure Clarence Thomas's appointment to the Supreme Court despite allegations of sexual harassment against Thomas. This marked the beginning of Leo's mission to reshape the judiciary. He began working for the Federalist Society in Washington, DC, later serving as its vice president for 25 years and currently co-chairing its board of directors.

Leo's Catholic faith profoundly shapes his worldview and political activities. He is a member of Opus Dei, a secretive Catholic organization dedicated to promoting Christian values in daily life. He belongs to the Sovereign Military Order of Malta, a chivalric order that traces its origins to the medieval Crusades. In her 2020 book *The Power Worshippers: Inside the Dangerous Rise of Religious Nationalism*,[71] journalist Katherine Stewart argues that Leo's project is fundamentally anti-democratic, writing:

> Leo's primary conviction is that democracy will not deliver the kind of conservative values-based government that he believes America must have. He is therefore committed to building an oligarchy of the religious and the wealthy.

Leo has received numerous accolades, including the Catholic Information Center's John Paul II New Evangelization Award in 2022 and an honorary doctorate from Benedictine College in 2023. In high-profile speeches at these events, however, his rhetoric has drawn criticism for its confrontational tone: he has warned graduates and Catholic audiences about "barbarians, secularists and [progressive] bigots" whom he accuses of trying to "threaten and delegitimize" Catholics and drive them into "professional and social exile," even comparing "current-day bigots" to the Ku Klux Klan.[72]

To finance his political projects, Leo has helped build a network of tax-exempt organizations. In 2021, billionaire Barre Seid transferred ownership of his electronics company, Tripp Lite, to Leo's Marble Freedom Trust, which then sold the firm for about 1.65 billion dollars—an arrangement news reports describe as the largest known single donation to a politically oriented nonprofit, structured to avoid capital-gains taxes. Leo has also been involved in for-profit entities such as CRC Advisors and BH Group, which watchdogs say have received tens of millions of dollars from Leo-aligned nonprofits since around 2016. His network is now under investigation by the DC attorney general. In 2024 the Senate Judiciary Committee subpoenaed Leo in its Supreme Court ethics probe into undisclosed gifts to justices—a subpoena he has refused to honor.[73]

Leo's influence on the federal judiciary has been widely described as transformative: for years he worked to build a conservative Supreme Court majority that would be willing to overturn *Roe v. Wade*, the 1973 decision recognizing a constitutional right to abortion. In June 2022, the Court did so in *Dobbs v. Jackson Women's Health Organization*, upholding Mississippi's 15-week abortion ban in a 6–3 judgment and, by a 5–4 vote, formally overruling *Roe* and *Planned Parenthood v. Casey*. In 2024, the Court's ruling in *Trump v. United States* further reshaped constitutional doctrine by recognizing absolute immunity for a president's core constitutional acts and at least presumptive immunity for other official acts, a decision made possible by the same conservative supermajority that Leo helped to assemble.[74]

The Heritage Foundation and Project 2025

The Heritage Foundation, a tax-exempt conservative think tank based in Washington, DC, has been a driving force behind Republican policymaking since its founding in 1973. One of its early financiers, Coors Brewing executive Joseph Coors, was described by his own brother as having politics "a little bit right of Attila the Hun." Heritage provided a detailed blueprint for Ronald Reagan's presidency in its 1981 manifesto *Mandate for Leadership*, and contemporaneous accounts suggest that roughly 60% of its 2,000 recommendations were implemented or initiated during Reagan's first year in office. Decades later, in preparation for Donald Trump's first term, the foundation helped vet potential appointees, and by mid-2018 more than 60 Heritage staff and alumni were serving in administration posts, reflecting the group's continuing influence over Republican governing agendas.[75]

In April 2023, the Heritage Foundation unveiled Project 2025, a roughly 920-page *Mandate for Leadership* meant as a governing blueprint for the next conservative president. Backed in part by funding from Leonard Leo's network and Koch-aligned donors to Project 2025 partner organizations, the initiative called for sweeping consolidation of presidential control over the executive branch, radical policy changes across dozens of agencies, and the recruitment and training of up to 20,000 vetted conservatives for key political appointments. Critics described Project 2025 as an authoritarian, Christian nationalist playbook that would erode democratic checks and balances and curtail protections for women, LGBTQ people, and people of color. Although Trump's campaign for a second term as president formally disavowed the project, by early 2026, independent trackers estimated that roughly half of Project 2025's recommended executive-branch actions had already been put into motion. [76] [77]

The Rejection of Science

The General Social Survey (GSS), funded by the National Science Foundation, has tracked Americans' attitudes toward science and other institutions since the early 1970s. An analysis by sociologist Gordon Gauchat found that while overall trust in the scientific community has remained relatively stable, it has declined markedly among conservatives and frequent churchgoers since the late 1970s and the Reagan era. Gauchat also reports that better-educated conservatives are now more likely than less-educated conservatives to distrust science, which he interprets as reflecting a view of science as aligned with liberal elites and government regulation rather than with common sense or religious tradition. [78]

The Polarization of America

With the metamorphosis of the Republican Party and the broader polarization of US politics, the country has become increasingly divided. Surveys by the Pew Research Center show that since the late 1960s, Americans' trust in the federal government has eroded while partisan attitudes have grown more radical and hostile. In the late 1960s, during the Johnson and early Nixon years, roughly two-thirds of Americans said they trusted the federal government to do the right thing almost always or most of the time, but by the mid-1970s, after the Vietnam War and Watergate at the end of the Nixon era, that figure had fallen to about 36%. Trust declined further over the following decade, reaching roughly a quarter of the public by the late 1970s and early 1980s around the Carter–Reagan transition, and since the mid-2000s—including during Donald Trump's first presidency—it has generally hovered between about 20% and 25%, marking the longest sustained period of low trust in at least half a century. [79]

A comparison of congressional voting patterns between the 1971–1972 and 2021–2022 sessions shows that while Democratic members of Congress have become somewhat more liberal, Republican lawmakers have shifted much further to the right, contributing disproportionately to the rise in polarization. By 2019, during Donald Trump's presidency, 79% of Democrats and 83% of Republicans rated members of the opposing party coldly on a 0-to-100 "feeling thermometer," and majorities in each party—55% of Republicans and 47% of Democrats—said members of the other party were more immoral than other Americans, with Republicans also especially likely to describe Democrats as more close-minded, dishonest, and unpatriotic. [80]

Despite these divisions, many Americans continue to see their country as exceptional. A 2021 survey by the Survey Center on American Life found that about six in ten Americans (62%) said they were extremely or very proud to be American, and other polling shows that large shares, especially among religious conservatives, believe God has granted the United States a special role in history. At the same time, the same research team reported that roughly four in ten Republicans (39%), compared with a much smaller share of Democrats, agreed that "if elected leaders will not protect America, the people must do it themselves, even if it requires violent actions," raising concerns about openness to political violence.

During Donald Trump's first term, international perceptions of the United States deteriorated to historic or near-historic lows in many allied countries. In 2020, favorable views of the US fell to around a quarter of the public in some European nations, while 59% in South Korea expressed a positive view.

After Joe Biden's inauguration, Pew found a sharp rebound: across 16 surveyed countries, the median share with a favorable view of the US jumped from 34% in 2020 to 62% in 2021, with increases of at least 25 percentage points in several key European allies and favorable views in South Korea climbing to 77%. [81]

Pew's 2025 Global Attitudes Survey, conducted after Trump's return to office, again found widespread skepticism: across 24 countries, a median of just 34 percent expressed confidence in Trump to do the right thing in world affairs, and overall views of the United States had declined in 15 countries compared with the previous year. [82]

The Role of Propaganda and Media

A central element of this transformation was the process of "othering" — portraying political opponents, minority groups, or dissenting voices as alien, immoral, or dangerous. By framing these groups as fundamentally different from "true" Americans, political leaders can justify increasingly radical positions and treat opponents as threats rather than fellow citizens, a pattern that research links to dehumanization and a greater willingness to condone political violence. While othering is not unique to any one era or party, the modern Republican Party's evolution shows how contemporary movements can harness this tactic to rally an in-group, consolidate power, and narrow the space for genuine democratic debate.

As the Republican Party transformed, its use of propaganda became increasingly deliberate. Leaders and allied media figures deployed repetitive messaging to cultivate an "us versus them" mentality, casting political opponents and dissenting groups as existential threats rather than legitimate rivals. Through carefully crafted slogans, emotionally charged narratives, and visual imagery at rallies, on television, and across digital platforms, political operatives shaped public opinion using techniques reminiscent of twentieth-century propaganda campaigns. These methods—grounded in repetition, fear and grievance appeals, and the gradual dehumanization of opponents—helped redefine the party's identity in the eyes of its supporters and narrowed the range of views considered acceptable within its ranks.

In this media ecosystem, outlets such as Fox News, nationally syndicated conservative talk-radio programs, and partisan online platforms have played a crucial role in amplifying these narratives. Fox hosts like Sean Hannity and Tucker Carlson, together with talk-radio figures such as Rush Limbaugh, built loyal audiences by offering a steady stream of tightly framed stories in which conservative viewpoints are constantly reinforced while opposing perspectives are dismissed as biased, un-American, or dangerous. Studies of this style of programming describe how selective story choice, emotionally charged monologues, and repetitive catchphrases create an echo chamber that hardens partisan identities and grievances. More recently, social media algorithms have intensified this dynamic by rewarding outrage and sensational content, increasing exposure to polarizing messages and heightening negative feelings toward political opponents. [83]

Exploiting the Dynamics of Conformity

The party's metamorphosis relied as much on psychological dynamics as on formal policy shifts. Decades of social-psychology research show that group processes such as conformity, diffusion of responsibility, and the dehumanization of perceived outsiders can strongly shape individual judgment and behavior. As Republican discourse grew more polarized, many activists and elected officials increasingly internalized a collective identity defined by opposition to perceived external threats, from political opponents to cultural "enemies." In this environment, dissent was discouraged and loyalty to the group became paramount, while social and professional pressures nudged those who might once have resisted toward silence or accommodation. Over time, even long-standing party veterans adopted rhetoric and tactics that earlier generations of conservatives would have regarded as fringe.

The Role of Think Tanks and Action Committees

The transformation was further reinforced by conservative think tanks, political action committees, and well-funded donor networks that helped redefine the party's ideology and messaging. Figures such as Charles and David Koch, together with organizations like the Heritage Foundation and the Federalist Society, promoted an agenda centered on aggressive free-market policies, hardline positions on immigration and crime, and deep skepticism of federal regulation and social welfare programs. By channeling substantial financial resources into advocacy groups, academic centers, judicial-selection efforts, and partisan media, these networks ensured that their ideological vision was not only amplified to conservative audiences but also embedded in the party's governing infrastructure.

The Emergence of a Cult of Personality

As institutional guardrails weakened and ideological litmus tests hardened, the party also became increasingly organized around personal loyalty to a single leader rather than to a shared program or set of principles. Rhetoric that once centered on abstract conservative values shifted toward defending, excusing, or celebrating the leader's words and actions, even when they broke with long-standing norms or earlier Republican positions. Public displays of allegiance at rallies, constant repetition of favored slogans, and social penalties for critics within the party all contributed to an atmosphere in which questioning the leader was equated with betraying the movement itself. This personalization of politics created fertile ground for the kind of cult dynamics explored in the next chapter, where devotion to Donald Trump would come to override traditional conservative commitments and further erode the space for dissent.

Ethical Reflections and Contemporary Implications

The Republican Party's metamorphosis highlights the ethical risks of political movements that prioritize group cohesion over individual critical thinking. Suppressing dissent and engineering political narratives to promote a singular view can produce an ideological homogeneity that stifles democratic debate. The deliberate use of propaganda and psychological techniques to shape political identity raises difficult questions about where legitimate persuasion ends and manipulation begins.

In today's digital age, traditional campaign posters and televised speeches have given way to algorithm-driven messaging and social-media echo chambers. Processes of conformity, "othering," and the cult of personality now unfold on platforms capable of influencing millions of people in real time. Reflecting on the Republican Party's transformation, we must ask: How can societies prevent political persuasion from crossing the line into manipulation? What safeguards—legal, institutional, and cultural—are needed to preserve open, democratic discourse in an era of advanced psychological and digital tactics?

Conclusion

The polarization of America, driven in part by the Republican Party's transformation, underscores a profound shift in how political groups mobilize support and shape ideology. By employing propaganda, exploiting group dynamics, and fostering intense "othering", the party has reshaped itself and influenced the broader political climate.

As we explore the evolution of mind control techniques, the Republican Party's case serves as both a cautionary tale and a lens for understanding modern political communication. These processes are critical for historians, political scientists, and citizens who value democratic debate and individual autonomy. In subsequent chapters, we will examine how similar principles have been applied in chemical, surgical, and digital contexts, each presenting unique ethical challenges and implications for the future of human freedom.

Classic experiments in social psychology help explain how ordinary people can be drawn into extreme rhetoric or actions when group pressures are strong. In Solomon Asch's conformity studies, most participants knowingly gave a wrong answer at least once rather than break with the group's consensus. Stanley Milgram's obedience experiments showed that many subjects were willing to administer what they believed were dangerous electric shocks when instructed by an authority figure, even while visibly distressed. The Stanford Prison Experiment, led by Philip Zimbardo, suggested that people can quickly internalize roles and act harshly toward an out-group once norms and expectations shift in that direction, although the study's methods and ethics have since been heavily criticized. Together, these findings underscore how pressures for conformity, obedience to leaders, and the dehumanization of perceived outsiders can make it easier for political movements to normalize once-fringe ideas.

CHAPTER 7 – DONALD TRUMP AND THE PHENOMINON OF CULTS

Introduction

In modern political history, few figures have exemplified the intersection of populist rhetoric, media manipulation, and cult-like group dynamics as vividly as Donald Trump. Despite being twice impeached during his first term and found guilty of sexual assault and corporate malfeasance, Trump achieved a return to the White House. This chapter examines how Trump's rise and the movement he fostered—the Make America Great Again (MAGA) phenomenon—mirror historical examples of mind control and cult formation. We explore the techniques he employed to shape public perception, mobilize supporters, and create an environment where loyalty is prized above critical thought.

The Rise of a Populist Icon

Trump's False Narrative

Donald Trump's first serious flirtation with national politics came in 2000, when he briefly sought the Reform Party's presidential nomination before withdrawing early in the primary season. He later registered as a Democrat in 2001, then re-registered as a Republican in 2009, positioning himself as a businessman outsider rather than a career politician. In June 2015, Trump formally announced his candidacy for the presidency as a Republican.

By the time of the 2016 election, public trust in the federal government had fallen to historic lows, with confidence especially weak among Republicans. Trump, who had never held elected office, capitalized on this disillusionment by promising to "drain the swamp" and dismantle what he labeled the "deep state." After a series of primary victories, including a decisive win in Indiana in May 2016, he became the presumptive Republican nominee. In July 2016, Trump selected Indiana Governor Mike Pence as his vice-presidential running mate, while Hillary Clinton and Tim Kaine headed the Democratic ticket.

Trump portrayed himself as a "stable genius" and a self-made billionaire, but publicly available evidence casts doubt on both claims. In 2016, his longtime lawyer Michael Cohen testified that he contacted the schools Trump had attended to threaten legal action if they released his academic records. Trump has repeatedly asserted that he graduated at the top of his class from the Wharton School of the University of Pennsylvania, yet his name does not appear on the university's published honors lists or on the Dean's List in student records from that period, suggesting that his academic performance was far more ordinary than he claims.

Trump's business acumen is similarly open to question. Over several decades he has presided over a string of high-profile bankruptcies and failed ventures. He filed for Chapter 11 bankruptcy protection for the Trump Taj Mahal casino in 1991, for the Trump Plaza, Trump Castle, and the Plaza Hotel in 1992, for Trump Hotels and Casino Resorts in 2004, and for Trump

Entertainment Resorts in 2009. Other ventures, including Trump University, Trump Vodka, Trump Mortgage LLC, and Trump Steaks, were eventually shuttered amid lawsuits, regulatory investigations, or persistent financial losses.

Trump's legal troubles are extensive. In 1973, the US Department of Justice's Civil Rights Division sued Trump Management, Donald Trump, and his father, Fred Trump Sr., alleging systematic violations of the Fair Housing Act by discriminating against Black renters in their New York apartment buildings covering roughly 14,000 apartments. In the 1990s, New Jersey regulators fined Trump's Atlantic City casinos for regulatory violations, including a 40,000-dollar penalty against Trump Plaza Hotel & Casino for employing unlicensed accountants. Trump's Taj Mahal casino also broke federal anti-money-laundering rules 106 times in its first 18 months of operation, resulting in a $477,700 fine in 1998 for Bank Secrecy Act violations. By the time Trump was elected president, a USA Today analysis found that he and his business entities had been involved in more than 3,500 lawsuits in federal and state courts. [84]

Five Rules to Live By

Trump's early mentor, the infamous lawyer Roy Cohn, modeled a ruthless style that biographers have distilled into a set of informal "rules" that Trump appears to have followed throughout his career. These include:

    1.   Never admit you're wrong; always project strength and confidence

2.   When challenged, go on the offensive and shift attention to your opponent

3.   Respond to criticism with even greater force

4.   Avoid conceding defeat or settling if you can instead prolong the fight

5.   Work relentlessly to define the narrative about yourself, regardless of the underlying facts. [85]

Entry into Politics

Upon becoming a serious Republican presidential contender in 2015–2016, Trump's preemptive sloganeering and grievance messaging intensified. He repeatedly claimed that the 2016 election was being "rigged," alleging without evidence that there would be widespread voter fraud, including illegal voting by non-citizens and people casting ballots more than once. His narcissism was captured in a now-famous remark at a January 2016 rally in Sioux Center, Iowa: "I could stand in the middle of Fifth Avenue and shoot somebody, and I wouldn't lose any voters." [86]

At some Trump rallies, supporters shouted "Lügenpresse" ("lying press"), a term popularized as a propaganda slur in Nazi Germany and revived by far-right movements to attack the media. Trump demonized the Democratic Party by branding his opponent "crooked Hillary" over her use of a private email server as Secretary of State and encouraging crowds to chant "lock her up." He also disparaged Republican Senator John McCain's war record, saying of prisoners of war, "He's a war hero because he was captured. I like people who weren't captured." Trump stoked anti-immigrant sentiment by promising to build a wall on the southern border and, in his June 2015 announcement speech, claiming that Mexican migrants were "bringing drugs. They're bringing crime. They're rapists." He likewise fueled anti-Muslim hostility by calling for a "total and complete shutdown of Muslims entering the United States until our country's representatives can figure out what is going on."[87]

The Presidential Debate and Election Outcome

Two weeks before the 2016 presidential election, during the final presidential debate, Trump refused to commit to accepting the result if Hillary Clinton won, saying he would "keep you in suspense." He ultimately secured the presidency with an Electoral College tally of 304 votes to Clinton's 227, but lost the national popular vote by about 2.8 million ballots, which he treated as a blemish on his victory.[88] Trump responded by asserting, without evidence, that Clinton had benefited from "millions" of illegal votes and alleging "serious voter fraud in Virginia, New Hampshire and California."[89] He then attacked the press for failing to validate these claims, accusing the media of "serious bias" and calling that failure a "big problem."

Dehumanization of Opponents

The dehumanization of Democrats became a recurring theme in Trump-era rhetoric, both from the president and his allies. In a February 2017 interview on Fox News's *Hannity*, broadcast shortly after Trump's inauguration, his son Eric Trump described Democratic critics by saying, "To me, they're not even people."[90] In May 2020, Trump amplified this sentiment by sharing a video from a supporter at a rally declaring, "The only good Democrat is a dead Democrat," adding his own comment of approval when he posted it.

Trump's Capacity for Falsehoods

Trump's propensity for falsehoods has been meticulously documented. According to The Washington Post Fact Checker database, he made 30,573 false or misleading claims during his four years in office, with the pace of untruths accelerating sharply in the run-up to the 2020 election.[91] The topics that generated the highest volumes of false or misleading statements included the coronavirus pandemic, health care, immigration, the economy, the Russia investigation, jobs, and the Ukraine probe.

The Trump Rallies

The Prelude

Trump's rallies, while lacking the grandeur of Hitler's Nuremberg events, were similarly stage-managed to captivate and unify attendees. Social psychologists Stephen Reicher and Alexander Haslam analyzed Trump's 2016 rallies in Ohio in their *Scientific American* article "The Politics of Hope: Donald Trump as an Entrepreneur of Identity," describing them as "identity festivals" that embodied a politics of hope.[92]

The rallies began long before Trump's arrival, with security procedures heightening the sense of importance. Attendees queued and passed through metal detectors, while security agents and rally staff scanned the crowd, making eye contact, watching for intruders, and encouraging attendees themselves to monitor for "subversives" or people showing insufficient enthusiasm. Well before the candidate appeared, crowds were led in rhythmic chants—"Trump! Trump! Trump!"—and instructed how to respond to protesters, reinforcing a sense that they were under threat from external and internal enemies.

The press corps, confined to a pen that was both visible to the crowd and set apart from it, became a ritualized target of Trump's attacks. He regularly pointed to the journalists as representatives of a hostile establishment, branding them "dishonest" and encouraging boos and harassment, which in turn dramatized the division between the loyal "in-group" and its supposed enemies.

The Stereotyped Performance

Trump's speeches followed a consistent formula. He opened by praising the crowd size, flattering attendees with lines like, "Wow. Whoa. That is some group of people. Thousands… This is beyond anybody's expectations. There's been no crowd like this," using the crowd itself as proof of his charismatic appeal and their collective power. He then declared that the American dream was failing because of external enemies (such as China and Mexico, said to be "stealing" jobs) and internal enemies (Obama, Clinton, and even Republican rivals), casting his supporters as a besieged "people" betrayed by corrupt elites.

Trump positioned himself as the savior outsider who could "make a deal" on their behalf. Amid chants of "We want Trump! We want Trump!" he promised, "If I get elected president, I will bring it back bigger, and better, and stronger than ever before, and we will make America great again," repeatedly invoking "we" to foster a sense of shared destiny and collective agency. Once in office, he continued to hold campaign-style rallies at airports, often addressing the crowd with the backdrop of Air Force One— a theatrical *deus ex machina* entrance that visually fused the leader, the state, and the movement, and echoed Hitler's Nuremburg Rallies.

The Clichés

Trump relied on simple, endlessly repeated slogans to indoctrinate and mobilize his audience, echoing authoritarian propaganda principles that stress memorability over truth. His cornerstone phrase was "Make America Great Again," reinforced by a repertoire of catch-phrases such as "Drain the Swamp," "Fake News," "America First," "Build the Wall," "Promises Made, Promises Kept," and "Lock Her Up," which condensed complex grievances into emotionally charged clichés.

Trump's Scorn for his Followers

While Trump's supporters were captivated by him, his feelings toward them were far from reciprocal. Howard Stern, a former friend, remarked on his radio show, "The people who are voting for Trump for the most part ... he wouldn't even let them in a fucking hotel. He'd be disgusted by them."[93] Michael Cohen, Trump's former attorney, revealed that Trump viewed his supporters as "low class" and "pawns to be used for his benefit."[94] Olivia Troye, a former national security advisor, reported that during the pandemic Trump commented, "You know, maybe this virus is a good thing. I don't have to shake hands with people anymore. I don't have to shake hands with those disgusting people."[95]

The Murdoch Factor

Trump excelled at self-promotion and propaganda, establishing a symbiotic relationship with the Murdoch press. He watched Fox News for hours daily, often tweeting in response to their stories. Emails released in 2022 revealed that the White House instructed Sean Hannity on what to say in his broadcasts.[96] Following his 2024 election, Trump appointed several Fox journalists to cabinet positions.

Social Media

Social media platforms, particularly Twitter and Facebook, were also key propaganda tools. During his first presidency, Trump tweeted 34,000 times and amassed 88 million followers, ranking sixth in popularity.[97] He used Twitter to attack political enemies, with *The New York Times* reporting that he targeted approximately 850 individuals in more than 6,000 tweets.[98]

Trump also retweeted content with racist or anti-Semitic themes and used Twitter to dismiss officials, including Rex Tillerson, Kirstjen Nielsen, and James Comey, often leaking dismissal letters to the media.

Is Trump a Fascist

Political theorists generally define fascism not by a single slogan but by a cluster of recurring traits: the creation of a mythic national "people" under threat from internal and external enemies; systematic lying and propaganda that delegitimize independent media and expertise; glorification of a strongman leader who claims to embody the nation's will; and a readiness to use or condone violence and legal transgression to crush opponents and overturn democratic constraints. In Trump's case, these traits are evident in his ethno-nationalist rhetoric and scapegoating of immigrants and minorities, his relentless use of falsehoods and "fake news" attacks to corrode shared reality, his personality cult and insistence that "only I can fix it," and his efforts to subvert elections, pardon political violence, and wield state power vindictively against perceived enemies. [99]

In short, when judged against these criteria, Trump's politics exhibit many of the core features scholars associate with fascism—distorted by his own peculiar combination of being vain, venal, vacuous and vicious:

- Vain – Obsessed with crowd size, television ratings, and personal adulation, he repeatedly claims "nobody's ever seen anything like this," prioritizing his image and ego over policy substance or institutional norms.

- Venal – He treats public office as a vehicle for personal and family enrichment, from continued business entanglements and foreign patronage at his properties to using the Justice Department to protect allies and punish critics.

- Vacuous – His rhetoric is dominated by slogans and superlatives rather than coherent programs, with major issues—from health care to foreign policy—reduced to vague promises that "we'll see what happens" or "we'll have something terrific."

- Vicious – He routinely dehumanizes opponents and vulnerable groups, encourages violence at rallies, demands prosecutions of rivals, excuses or pardons political thuggery, and incited an attempt to overturn an election that resulted in multiple deaths and scores of injuries.

Trump's Techniques of Political Mind Control

Propaganda and Repetition

Trump relies heavily on repetitive, simplified messaging. Slogans such as "Make America Great Again," "Drain the Swamp," "Build the Wall," and "Fake News" are designed to be short, memorable, and emotionally charged. They are repeated across rallies, tweets, hats, and banners. They turn complex grievances into binary choices between a corrupt "them" and a virtuous "us." This constant repetition creates a shared language for his movement and avoids more nuanced policy discussion.

Exploiting group dynamics and "othering"

His rhetoric systematically divides the world into loyal "patriots" and dangerous "others"—immigrants, Muslims, urban liberals, journalists, "elites"—who are blamed for the nation's decline. By constantly framing his followers as a besieged in-group whose identity and status are under attack, Trump activates powerful group-based emotions of fear, anger, and resentment, which social psychologists have shown can suppress critical judgment and increase conformity to a leader's cues. This "othering" makes harsh policies and even political violence against opponents seem like justified self-defense.

Digital propaganda and social media

Where earlier demagogues depended on newspapers and radio, Trump exploits Twitter, Facebook, YouTube and allied media to broadcast unfiltered messages, attack enemies, and test slogans in real time.[100] Social-media algorithms, tuned to maximize engagement, tend to amplify the most polarizing content and foster echo chambers in which users mainly see reinforcing in-group messages and negative portrayals of perceived enemies, strengthening group identity and ideological homogeneity. These same tools enable micro-targeted political advertising and rapid coordination of supporters, making the psychological pull of his propaganda both more personalized and more pervasive.[101]

Looking to the Future

As political movements evolve, the lessons from Trump's era highlight the need for vigilance. Democratic societies must ensure that political persuasion does not devolve into manipulation or coercion. Achieving this requires greater transparency in political communication, robust regulation of digital platforms, and a renewed focus on fostering critical, independent thought among citizens.

Conclusion

Trump's transformation into a near-cult figure exemplifies how modern politics can exploit ancient mind control techniques. Through repetition, emotional appeals, and deliberate "othering." Trump and his MAGA movement have reshaped the US political landscape. This chapter has examined how these strategies mobilized a powerful constituency while presenting profound ethical challenges.

Understanding the dynamics of Trump's rise—and the parallels between his methods and historical propaganda—underscores the importance of safeguarding democratic discourse. In an era where technology enables rapid and far-reaching manipulation of public opinion, the lessons of this chapter serve as both a warning and a call to action.

CHAPTER 8 - MAGA, AND OTHER CULTS

Introduction

While political movements such as MAGA display many techniques of psychological manipulation, cults and other high-control groups offer even deeper insight into how charismatic leadership and social pressure can reshape individual behaviour. NXIVM—an organization that marketed itself as self-improvement training while concealing a web of coercive and exploitative practices—shows how such methods can extend far beyond formal politics. In this chapter, I examine how the psychological techniques used to build the MAGA movement parallel those employed in the genesis of non-political cults, using NXIVM as a central case study.

NXIVM – A Cult Beyond Politics

Inception

In 1998, Keith Raniere and Nancy Salzman founded NXIVM in New York's Capital District, presenting it as a company offering personal development seminars. [102] Raniere had previously run a business accused of operating as a pyramid scheme, while Salzman, a nurse with a background in psychiatric care, supplied clinical credibility and therapeutic language. Together they recruited followers through a multi-level marketing model and built their curriculum around "Rational Inquiry," a proprietary system Raniere claimed to have developed to unlock human potential.

Fortifying Cult Ideology Through Wealth and Fame

At its peak, NXIVM counted roughly 700 active members, including actors, lawyers, and the children of wealthy and influential families. In 2009, the Dalai Lama appeared at an event sponsored by Raniere, while Roger Stone—a long-time advisor to Donald Trump—briefly served as one of NXIVM's lobbyists, further lending the group an aura of legitimacy and influence. [103]

A secret society within NXIVM, known as DOS (derived from the Latin Dominus Obsequious Sororium, loosely interpreted as "master over slave"), epitomized its darkest practices. In DOS, women were branded with Raniere's initials, forced to submit written confessions, and coerced into providing nude photographs that could be used as blackmail, solidifying the group's grip through both psychological terror and control over personal identity. [104]

For many who joined, the presence of celebrities, wealthy patrons, and even revered spiritual figures made NXIVM feel less like a fringe experiment and more like a rare opportunity they would be foolish to pass up. Newcomers described the early courses as intense but exhilarating: rooms filled with well-dressed professionals trading stories of breakthroughs, instructors speaking in polished, therapeutic language, and constant assurances that they were part of a select group brave enough to "do the work" others avoided. In that environment, doubts were easy to dismiss. If billionaires, respected public figures, and seemingly empowered women all endorsed the program, how could anything truly sinister be hiding beneath the surface?

By the time subtle humiliations and alarming demands began to appear, many members had already reshaped their lives around the group—moving cities, investing savings, ending relationships with skeptical partners or parents. Those who were funneled into DOS often entered believing they were being invited into an elite sisterhood dedicated to radical growth, not a system of sexual servitude. Branding, starvation diets, and sleep deprivation did not arrive all at once; they emerged gradually, each step framed as a test of commitment or a chance to overcome "limitations." Survivors later recalled that the most frightening part was not a single overt act of cruelty, but the slow realization that they had been trained to call abuse "empowerment" and to protect the man orchestrating it, even from their own instincts.

The Psychological Tools Employed

Loyalty to a Charismatic Leader and Suppression of Dissent

Followers were indoctrinated to revere Raniere, whom they referred to as "Vanguard," and were taught to regard him as the smartest and most ethical man alive. Internal propaganda claimed that his "intellectual energy" could activate latent abilities in his followers and that his semen possessed supernatural properties capable of healing what he termed their "disintegrations." Many members did not encounter Raniere in person until they were deeply embedded in the organization, which only heightened his aura as a distant, mysterious figure who should be revered.

Within the organization, dissent was not tolerated. Critics or perceived "enemies" were targeted with psychological and, at times, physical punishment, and adversaries were discredited or intimidated through harassment, litigation, and related tactics.

Obedience and Readiness

Raniere's authority was effectively absolute. Members were required to perform degrading and demeaning tasks—such as licking water from a street puddle—without question, framed as "growth opportunities" or penances when they failed to meet his standards.

"Readiness drills" demanded immediate responses to text messages at all hours, with a failure by one member triggering punishment for the entire group, thereby enforcing collective discipline and unquestioning obedience.

Rituals, Symbols, and the Power of Community

NXIVM's rituals and symbols, including specialized handshakes, honorifics, and the secretive branding ceremonies within DOS, deepened members' psychological investment and signaled membership in an elite, enlightened community. These practices fostered a powerful sense of belonging, even as they bound individuals more tightly to the group's demands.

Othering"

A fundamental mechanism of control was "othering." Members were encouraged, and often pressured, to sever ties with friends and family who questioned NXIVM, gradually creating a dependency that rested almost entirely on the group. Distinctive jargon and esoteric concepts reinforced this separation, heightening a shared in-group identity while instilling distrust and contempt toward outside perspectives.

Deindividuation

Drawing parallels to research such as Zimbardo's Stanford Prison Experiment, Raniere and his lieutenants stripped members of individual identity and autonomy. In "Society of Protectors" sessions, male members assumed quasi-guard roles, humiliating female participants to reinforce rigid gender and power hierarchies. DOS members were not only branded but were also forced to wear dog collars and, in some cases, confined to dog cages, further eroding their sense of self and normalizing degradation.

Depletion

Physical and psychological depletion further undermined resistance. Members were subjected to severe sleep deprivation, chronic under-nourishment, and extreme physical challenges such as ice-cold showers and late-night or winter walks, often framed as character-building exercises. These conditions produced exhaustion and cognitive impairment, making individuals more susceptible to manipulation and less able to question instructions or consider leaving.

The Cult's Demise

In 2018, following revelations by whistleblowers and investigative reporting, Keith Raniere and other key members were arrested on charges including racketeering, sex trafficking, forced labor, conspiracy, wire fraud, and related offenses. In 2019, Raniere was convicted on all counts and, in 2020, was sentenced to 120 years in federal prison. [105]

Even after his incarceration, Raniere continued to exert influence over a small cadre of devotees, urging them to defend him publicly and recruit others. Many former members have since sought professional counseling to address the profound psychological effects of their involvement and to rebuild relationships damaged by years of isolation and control.

The MAGA Movement

Inception

In November 2012, ahead of his 2016 presidential bid, Trump filed a trademark application for the slogan "Make America Great Again" (MAGA) for political use, and it was registered in July 2015 once he began using it in his campaign. [106] Around the same period, his campaign also arranged to acquire rights from a Dallas businessman who had separately sought to trademark the phrase, reimbursing his costs via a charitable donation so that Trump's team could consolidate control over the slogan's commercial and political use.

MAGA was built on the premise that the United States had squandered its international standing, undermined from within by immigration and multiculturalism and from without by globalization and trade deals seen as unfair.[107] The movement promoted sharply reduced immigration, including measures targeting Muslim-majority countries. It embraced economic protectionism and border fortification, most famously Trump's proposal to curb illegal immigration from Mexico by constructing a wall along the 3,145-kilometre border.

Within the Republican Party, Trump increasingly cast himself as a kingmaker, using endorsements, rallies, and public threats to elevate MAGA-aligned contenders in state and federal races and to punish critics.

As the MAGA movement evolved, it absorbed and amplified sexist and anti-LGBTQ+ rhetoric, sharpened its attacks on "woke" culture, and treated mainstream news organizations as enemies of the people, while drawing significant support from parts of the Murdoch media ecosystem. It promoted an increasingly nationalist and isolationist foreign policy stance.

After the 2020 election, MAGA followers rallied around Trump's baseless claim that the presidency had been stolen through massive fraud. MAGA-aligned activists and groups were highly visible among those who stormed the US Capitol on January 6, 2021, in an effort to overturn the result.[108]

Political Movement or Cult?

Although MAGA presents itself as a political movement, it functions in many respects like a cult and is arguably one of the largest such formations in contemporary US politics.[109]

In January 2024, PLOS ONE published a nationwide survey of MAGA Republicans led by Garen J. Wintemute and colleagues from the Violence Prevention Research Program at the University of California, Davis. [110] In their analysis, a "MAGA Republican" was defined as a Republican who voted for Trump in 2020 and strongly agreed with the statement, "The 2020 election was stolen from Donald Trump, and Joe Biden is an illegitimate president." Using this definition, they estimated that MAGA Republicans comprised about one-third of all Republicans and approximately 15% of the adult US population—roughly 39 million people.

MAGA followers in the study were predominantly white, male, Christian, and older, with a substantial share retired and about half located in the middle-income range. They were heavily involved in conservative advocacy organizations: roughly 85% reported membership in gun-rights groups, and about half said they belonged to anti-abortion ("pro-life") organizations. Most endorsed conspiracy theories. Some 82% agreed with a version of the "Great Replacement" claim that native-born white Americans are being replaced by immigrants, and a majority endorsed QAnon-style beliefs that US institutions are controlled by a cabal of elites involved in child sexual exploitation.

Regarding political violence, roughly nine in ten MAGA Republicans agreed that there is a "serious threat to our democracy," and more than three-quarters thought that "in the next few years, there will be civil war in the United States." When asked about the justification of "force or violence" to advance important political objectives, large majorities considered violence at least sometimes justified for at least one of 17 specific aims, including "returning Donald Trump to the presidency this year," "preserving an American way of life based on Western European traditions," and "stopping police violence."

The authors emphasized that while MAGA Republicans were significantly more likely than other groups to endorse political violence in principle, they were not more likely to say they themselves were very or completely willing to engage in such violence—an important distinction between support and personal intent.

Even so, Wintemute and colleagues warned that their findings show "our country is in grave danger," given that one of the two major parties contains a large faction that denies the legitimacy of national elections and is notably more supportive of political violence than other Americans.

The Cult-Like Dynamics of the MAGA Movement

Loyalty to a Charismatic Leader and the Suppression of Dissent

Loyalty to Trump is the cornerstone of the MAGA movement. Supporters—whether political aspirants or ordinary voters—are expected to adopt the leader's perspective with minimal questioning, and public skepticism or dissent is often met with accusations of disloyalty, betrayal, or even unpatriotic behavior. This enforced conformity closely mirrors high-control groups in which critical thinking is discouraged, and collective identity is elevated above individual discernment.

Rituals, Symbols, and the Power of Community

Rallies, branded merchandise, and online forums function as the rituals and symbols of the MAGA movement. Red hats, flags, and slogans not only signal membership but foster a sense of belonging and shared purpose, turning political support into a lived identity.

Much like religious or ideological cults, the collective experiences—chants, call-and-response slogans, and large public demonstrations—strengthen group cohesion and deepen members' emotional investment in the movement.

When Donald Trump's former lawyer Michael Cohen derisively referred to him as "Von ShitzInPantz" in social-media posts, some Trump supporters responded by ironically embracing the insult, appearing at rallies in adult diapers and brandishing slogans such as "Real Men Wear Diapers" and "Diapers over Dems." This willingness to turn ridicule into a badge of honour illustrates a familiar cult dynamic in which humiliating or mocking labels from outsiders are reappropriated to demonstrate unconditional loyalty to the leader.

"Othering"

Many MAGA followers perceive themselves as a uniquely enlightened in-group whose leader—a former and aspiring president—embodies the authentic will of "real" Americans. The out-group is cast broadly, encompassing immigrants, mainstream media, cultural and political "elites," and especially Democrats and Republicans deemed insufficiently loyal to Trump. This sharp us-versus-them boundary reinforces psychological dependence on the leader and the movement, as external criticism becomes further evidence that outsiders are corrupt or dangerous.

Deindividuation

MAGA's deindividuation processes work by stripping away personal identity and replacing it with a collective identity centred on loyalty to Trump and the movement. Mass rallies feature repetitive chants such as "Make America Great Again," "USA," or "Lock her up," encouraging participants to merge their thoughts and emotions with the crowd, and rewarding those who echo the loudest and most transgressive messages. Visible markers—MAGA hats, flags, themed clothing, and vehicle decals—operate as constant, public affirmations of membership, making it easier for individuals to see themselves primarily as part of a unified, embattled group rather than as independent political actors.

Online platforms amplify this collective identity by creating spaces where dissenting opinions are filtered out or aggressively sanctioned. Algorithmic feeds and group norms prioritize content that reinforces pro-Trump narratives and conspiracy claims, further insulating members from alternative viewpoints and social circles. Fear-laden messages about existential threats to American values, demographic change, and "stolen" elections heighten emotional arousal, which can overshadow personal discernment and make individuals more likely to conform to the group's storyline.

Depletion and Emotional Manipulation

Rather than physically exhausting followers in the way that some closed cults do, the MAGA ecosystem relies heavily on emotional depletion and manipulation. A constant stream of outrage, fear-mongering about enemies "destroying America," and polarizing rhetoric keeps supporters in a state of heightened anxiety and grievance, encouraging them to see ordinary political setbacks as catastrophic betrayals. In this emotional climate, loyalty to Trump and the movement becomes both a coping mechanism and a moral duty, further entrenching the cult-like dynamics at the heart of MAGA.

Steven Hassan's BITE model (Behavior, Information, Thought, and Emotional control) offers one useful lens for understanding these dynamics. [111] In behavioral terms, MAGA's dense calendar of rallies, online engagement, and activist tasks structures the followers' time and social worlds around the movement; information control appears in its hostile framing of mainstream media, reliance on partisan outlets, and online ecosystems that filter out dissenting views. Thought control is reflected in binary "patriots versus traitors" rhetoric and loaded language that casts critics as enemies of "real America," while emotional control is maintained through constant cycles of outrage, fear, and grievance that keep supporters anxious, morally energized, and tightly bound to Trump as the only figure who can avert catastrophe. Critics of Hassan have argued that the BITE model can be applied too broadly and risks labeling many ordinary organizations as "cults," yet when focused on extremes of loyalty, information isolation, and apocalyptic emotion, it highlights why MAGA can plausibly be described as a high-control movement rather than just a conventional political faction.

Taken together, these patterns suggest that MAGA scores unusually high on all four BITE dimensions for a mainstream political tendency, making its psychology and social structure resemble a high-control cult more than a conventional electoral movement.

BITE domain	Key feature in Hassan's model	How the MAGA movement fits
Behavior	Regulates members' time, activities, and social environment (meetings, tasks, social circles).	Dense ecosystem of rallies, canvassing, social media activism, and themed community events that structure free time and social ties around MAGA identity.
Information	Filters information, discourages or demonizes external sources, promotes an "insider" narrative.	Mainstream outlets branded "fake news" or "enemy of the people," reliance on partisan and alternative media, and online groups where dissenting links or views are ridiculed or removed.
Thought	Encourages black-and-white thinking, loaded language, and doctrine over personal evaluation.	Binary framing of "patriots" versus "traitors," "real Americans" versus "globalists," plus slogans and conspiracy narratives that define loyalty as accepting Trump's version of events.
Emotional	Uses guilt, fear, and exaltation to maintain commitment and discourage questioning.	Persistent fear-messaging about invasion, demographic "replacement," stolen elections, and civil war, combined with euphoric rallies and a sense of heroic struggle that makes doubt feel like betrayal.

Ethical Implications and Contemporary Challenges

The techniques employed by Trump and his supporters raise significant ethical concerns. When political persuasion begins to resemble mind control or cult indoctrination, the space for open debate and independent thought diminishes. The manipulation of public sentiment through propaganda and digital echo chambers undermines the foundations of democratic discourse. [112]

The Role of Technology in Political Manipulation

Modern technology plays a dual role in political movements. While digital platforms have democratized information sharing, they have also provided powerful tools for manipulation. Social media algorithms amplify polarizing content, effectively "programming" public opinion. The ethical challenge lies in balancing the free flow of information with safeguards against coercive tactics. [113]

Conclusion

This chapter demonstrates how the political MAGA movement emulated the psychological techniques used by the powerful and notorious NXIVM cult to achieve its ends. Charismatic leadership, "othering," obedience, suppression of dissent, rituals and symbols, deindividualization, and depletion and emotional manipulation were used by both with good effect.

As political movements evolve, the lessons from Trump's era highlight the need for vigilance. Democratic societies must ensure that political persuasion does not devolve into manipulation or coercion. Achieving this requires greater transparency in political communication, robust regulation of digital platforms, and a renewed focus on fostering critical, independent thought among citizens.

CHAPTER 9 – ZERSETZUNG IN THE GERMAN DEMOCRATIC REPUBLIC

Introduction

The establishment of the German Democratic Republic (GDR) in 1949 marked the beginning of one of the world's most pervasive modern surveillance states. At the heart of this system was the Ministry for State Security (Ministerium für Staatssicherheit), better known as the Stasi, whose psychological warfare techniques were encapsulated in the term "Zersetzung" ("decomposition" or "undermining"). Unlike overt repression, Zersetzung aimed to subtly erode an individual's sense of self, destabilize relationships, and neutralize perceived enemies of the regime without leaving obvious traces of state violence. This chapter explores the origins, techniques, and enduring consequences of Zersetzung, highlighting lessons that resonate in an era of digital surveillance, targeted harassment, and psychological manipulation.

Historical Background

In October 1949, following the post-war division of Germany, the GDR was proclaimed under the leadership of Walter Ulbricht and the Socialist Unity Party (Sozialistische Einheitspartei Deutschlands, SED). While not formally incorporated into the Soviet Union, the GDR functioned as a loyal satellite state, closely aligned with Moscow in ideology, security policy, and economic planning.

In February 1950, the new regime created the Stasi, which quickly evolved into one of the most extensive and intrusive secret police organizations in history. [114] Ulbricht relied heavily on propaganda to legitimize the one-party state, saturating media and education with socialist doctrine while branding dissenters as "class enemies."

Open repression—including mass arrests, torture, and show trials—was used to terrorize the population and deter organized opposition. When workers' protests erupted in June 1953, Soviet tanks and GDR security forces crushed the uprising; thousands were arrested, and numerous protesters were imprisoned or executed in the aftermath. [115]

By May 1971, Ulbricht was replaced by Erich Honecker. Seeking greater international legitimacy and improved relations with West Germany, Honecker presided over a shift in the Stasi's methods from overt, highly visible terror to more covert and "scientific" forms of control.

Under Hinecker's leadership, techniques of Zersetzung that had been developed within the security apparatus were systematized and deployed more widely, aiming to break opponents psychologically while maintaining a façade of legality and normalcy.

The true extent of these paranoia-inducing operations only became clear after the fall of the Berlin Wall on 9 November 1989. When citizens gained access to Stasi files, they discovered that the agency had compiled records on roughly six million East Germans—more than one-third of the GDR's 1990 population of

16.3 million. This vast surveillance network was maintained by around 100,000 full-time employees and some 189,000 unofficial collaborators, many of them neighbors, colleagues, and even family members, ensuring that the reach of the secret police penetrated deeply into everyday life. [116]

Pervading Society

Zersetzung, developed and systematized within the Stasi's training institutions (including its Juristische Hochschule, or College of Law), was described as a form of "operational psychology" aimed at isolating individuals, inducing self-doubt, fear, and paranoia, and eroding resistance to state ideology. [117] It was labor-intensive: internal guidelines envisaged several operatives monitoring each targeted person or group, and thousands of citizens were subjected to sustained campaigns over many years.

The main components of Zersetzung included:

- Surveillance: Constant monitoring—both physical and electronic — collected information about a person's social, familial, and religious life, employment, political attitudes, sexual relationships, and everyday habits.

- Psychological warfare: Operatives sought to induce anxiety, depression, and paranoia by spreading rumours, gaslighting targets about what was "really" happening, and making them acutely aware they were being watched. Tactics included damaging property, tampering with vehicles, interfering with food and medicines, arranging inappropriate medical treatment, and issuing frivolous fines. Break-ins were staged not to steal but to rearrange furniture, alter clocks, or change alarms, slowly undermining victims' confidence in their own memory and sanity.

- Interference with personal affairs: The Stasi manipulated mail and communications, sending compromising—and sometimes fabricated—letters or photographs to spouses, relatives, or employers. Targets were transferred to remote or inconvenient workplaces, denied promotion, or blocked from professional advancement.

- Isolation and alienation: Victims were deliberately cut off from family, friends, and colleagues. At work, the Stasi might engineer sudden promotions or suspicious "favors" for the target, encouraging others to see them as informers or party loyalists and thereby deepening their social isolation.

Modern states, including the People's Republic of China, have developed extraordinarily powerful systems for collecting and analyzing data on their own populations. What made Zersetzung distinctive was not surveillance alone but the systematic, personalized use of that information to wage everyday psychological warfare against selected individuals and circles, with the aim of breaking opposition without visible terror. [118]

Psycho-social Implications

Erosion of Individual Autonomy

Zersetzung was not just about silencing opposition—it sought to fundamentally alter individuals. By systematically undermining confidence and destabilizing social networks, the regime stripped victims of autonomy and identity, often leaving long-lasting psychological scars.

This raises significant ethical questions about the boundaries of state power and the sanctity of individual thought, particularly when repression is designed to be invisible and plausibly deniable.pubmed.ncbi.nlm.nih+2

Lasting Social Impact

Zersetzung's legacy extends beyond the individuals it targeted, deeply affecting society as a whole. Fear of covert manipulation eroded trust among citizens and contributed to the persistent mistrust and political alienation observed in parts of the former GDR after reunification.

Efforts to address the harm have included public-education initiatives and memorial work, granting victims access to their surveillance files to understand how and why they were monitored, and various forms of rehabilitation and compensation such as restitution pensions and psychological counselling.[119]

Modern Echoes and Lessons

Digital Parallels

Although Zersetzung predates the digital era, its core principles have resurfaced in new technological guises. Digital surveillance, large-scale data collection, algorithm-driven content curation, and coordinated misinformation campaigns can isolate individuals, distort perceived reality, and weaponize social relationships in strikingly similar ways.

Recognizing these parallels is crucial for designing safeguards against technological abuse and preventing modern security practices from drifting into covert psychological warfare.

Safeguarding Autonomy in the Digital Age

The ethical dilemmas posed by Zersetzung point toward the need for contemporary protections of psychological and informational autonomy. Transparent data practices, robust privacy laws, effective oversight of security agencies, and a vigilant civil society are essential to limit covert manipulation and restore trust.

Just as Germany now draws on its own history of secret-police abuses to advocate for human rights and informational self-determination, modern societies must ensure that powerful technologies are deployed to empower citizens rather than to monitor, intimidate, or quietly break them. [120]

Conclusion

Zersetzung in the GDR exemplifies how a state can manipulate public behavior and individual identity. By systematically eroding trust and autonomy, the Stasi created an environment where dissent was nearly impossible. The lessons of this era serve as critical warnings for the present and future, especially as digital technologies provide new tools for surveillance and control.

As we delve deeper into the exploration of mind control—from chemical and surgical methods to algorithmic manipulation—it is vital to remain vigilant against the dangers of covert power. Only by understanding these dark chapters of history can societies build futures rooted in freedom and dignity.

Survivor Testimony: The Systematic Unraveling of Self

I was a political dissident in East Germany, and before long, I discovered that the state's methods were designed to dismantle who I was—from the inside out. It began with subtle disturbances that gradually morphed into a relentless assault. At first, I noticed small, inexplicable changes in my daily life: a misplaced letter here, an intercepted phone call there. Soon, strangers started paying unusual attention to my routines, and whispered rumors circulated, branding me an enemy of the state.

My personal space was invaded without warning. Objects in my home would be subtly rearranged, as if to remind me that nothing was safe or truly mine. Over time, the constant surveillance made every knock at the door, every unexpected glance from a passing stranger, a piercing reminder that I was being watched and manipulated. At work and among friends, fabricated stories began to spread. Former allies became skeptical, and I found myself increasingly isolated and mistrusted.

The psychological assault escalated slowly but deliberately. I was gaslighted through repeated, insidious lies and public smear campaigns, which gradually eroded the confidence I once had in my own memories and identity. I began to doubt not only my actions but my very thoughts. The pressure was so immense that it induced a state of chronic paranoia; I felt as though every aspect of my existence was under attack.

At the worst moments, despair overwhelmed me. The meticulous, invasive harassment left me disoriented, questioning where the state's influence ended and my own identity began. Although I eventually found support in a few trusted friends who helped me reclaim parts of my true self, the scars of those harrowing months remain with me. They stand as a constant testament to the power of Zersetzung—a methodical, psychological warfare that can strip away one's sense of self without ever raising a single physical blow.

PART 2 - THE CHEMICAL ASSAULT ON THE MIND

"Turn on, tune in, drop out."

Timothy Leary (1920 – 1996)

CHAPTER 10 – FROM STONE AGE TO STONED AGE

Introduction

Long before the advent of modern science and technology, humans discovered natural substances capable of altering perception, evoking powerful emotions, and sparking experiences they interpreted as spiritual revelations. These substances, initially explored out of curiosity or chance, evolved into tools for ritual, religious practice, and, at times, social control. This chapter examines the ancient roots of chemical influence over the mind, showing how entheogens (substances that induce mystical-type experiences) shaped early human cultures and how they continue to inform contemporary therapeutic practices and the potential for manipulation today.

Entheogens in Prehistoric Culture

The Spanish archaeologist Elisa Guerra-Doce has synthesized archaeological evidence for psychoactive substance use in prehistoric Europe, drawing on plant remains, chemical traces, and iconography to argue that altered states of consciousness were tightly woven into ritual and belief rather than casual entertainment.[121] Her work builds on a classic cross-cultural survey by anthropologist Erika Bourguignon, who analysed 488 ethnographic societies and found that about 90 percent incorporated altered states of consciousness into their fundamental belief systems. Guerra-Doce focused on four main lines of archaeological evidence: fossils of psychoactive plants, psychoactive alkaloids detected in skeletal material and artefacts, residues of alcoholic beverages, and depictions of drinking scenes or mood-altering plants in prehistoric art.

The earliest traces of such practices reach back many millennia. Evidence from Asia suggests the use of stimulating betel leaves thousands of years ago, while in northern Mexico and Texas hallucinogenic mescal beans associated with ritual contexts have been dated to the early Holocene, roughly 10,000–11,000 years ago. Because most finds of psychoactive plants and drink residues come from tombs and restricted ceremonial sites, researchers infer that access to these substances was tightly controlled and reserved for rites thought to enable communication with other realms, access to esoteric knowledge, or accompaniment of the dead into the afterlife.

In 1933, Lieutenant Charles Brenans of the French Foreign Legion drew attention to the Tassili n'Ajjer region in present-day Algeria, home to one of the world's richest concentrations of prehistoric rock art. [122] Over 15,000 carvings and paintings, some dating back up to 10,000–12,000 years, depict animals such as elephants and giraffes, human activities like hunting and dancing, and enigmatic "round-headed" figures that many interpret as shamans or masked ritual participants. Among these are images sometimes described as "mushroom figures"—humanoid forms with mushroom-like shapes sprouting from their bodies—which some scholars take as possible depictions of ritual use of psilocybin-producing fungi, although this interpretation remains contested.

Entheogens and Contemporary Religions

Entheogen Use in Hinduism

In ancient India, the Rig Veda—a foundational text of early Vedic religion, composed roughly three millennia ago—repeatedly praises Soma, a divine elixir said to confer immortality, inspire ecstatic insight, and forge a direct connection with the gods. The botanical identity of Soma remains disputed: candidates have included psychedelic mushrooms, ephedra, and other plants containing psychoactive alkaloids, but no single hypothesis commands universal agreement.

Entheogens in Christian Culture

Some researchers have argued that entheogens also left their mark on Christian art and symbolism. In their 2016 book The Psychedelic Gospels: The Secret History of Hallucinogens in Christianity, Jerry and Julie Brown point to images resembling the muscimol-containing Amanita muscaria mushroom in medieval churches across Europe and in present-day Turkey.[123] One of their most cited examples is a thirteenth-century fresco of the Temptation in the Garden of Eden in the Chapel of Plaincourault in France, which depicts the serpent coiled around a large mushroom-like form that the Browns interpret as Amanita muscaria rather than a conventional tree. They suggest that the biblical "tree of knowledge of good and evil" in Genesis may, in some contexts, have symbolized a visionary "tree" of psychedelic experience rather than an apple. They identify similar mushroom-like motifs in the Great Canterbury Psalter, mosaics at the Basilica of Aquileia, and frescos in churches such as Saint Martin de Vic, though these interpretations remain controversial among art historians.

The Effect of Entheogens

Two classic studies illustrate the potential of entheogens to occasion long-lasting spiritual and mystical-type experiences. In 1962, Walter Pahnke, a graduate student at Harvard Divinity School, conducted the Good Friday (Marsh Chapel) Experiment under the supervision of Timothy Leary and Richard Alpert.[124] Twenty Protestant divinity students took part in a double-blind study during a Good Friday service at Boston University's Marsh Chapel. Ten received psilocybin, and ten received an active placebo. Those in the psilocybin group reported profound mystical experiences closely matching criteria for "classic" mystical states, and follow-up reports suggested substantial positive changes in attitudes and behaviour at six months and beyond; decades later, surviving participants still described the event as one of the most meaningful in their spiritual lives.

In 2008, Roland Griffiths and colleagues at Johns Hopkins University published a 14-month follow-up of 36 hallucinogen-naïve, well-educated adults who had received psilocybin under supportive, controlled conditions.[125] At follow-up, 58–67 percent of participants rated the psilocybin-occasioned session as among the five most personally meaningful and spiritually significant experiences of their lives, and about two-thirds reported persistent increases in well-being or life satisfaction that they attributed to the experience.

Entheogens' transformative effects are not confined to explicitly religious settings. A 2023 study by Canadian psychologists Kevin O. St. Arnaud and Donald Sharpe found that entheogen use was strongly associated with self-transcendent states, feelings of awe and connectedness, and indicators of psychospiritual growth, even among non-religious users. [126] Anecdotal accounts from atheists and agnostics similarly describe experiences with substances such as ayahuasca or psilocybin as among the most meaningful events of their lives, often framed in terms of personal insight, emotional healing, or a renewed sense of purpose rather than traditional religious belief. [127]

Contemporary Psychedelic Use

In the early 20th century, stories of nocturnal mushroom rituals among Indigenous communities in Mexico began to filter into Western anthropology. In 1936, engineer and amateur ethnographer Robert Weitlaner collected specimens of mushrooms used in a divinatory ceremony in northeastern Oaxaca; because the material was poorly preserved, they were initially assigned only to the genus *Panaeolus*, which is now recognised to include psilocybin-containing species. [128] Two years later, members of Weitlaner's circle became some of the first non-Indigenous outsiders in centuries to witness such a ritual. In the ceremonies they described, the male shaman was often the sole consumer of the mushrooms, directing his divinatory pronouncements to spirits of nature, celestial bodies, and Catholic saints. [129]

Robert Gordon Wasson

Robert Gordon Wasson, a New York banker and amateur mycologist, gained worldwide recognition for his work with Mexican "magic mushrooms," helped by his own talent for self-promotion, the backing of media allies such as Henry Luce of Time Inc., and, indirectly, by the interest of US intelligence agencies. Archival correspondence shows that Wasson exchanged letters over many years with CIA officer Paul Charles Blum, and declassified documents link his fieldwork to MKUltra funding structures;[130] the CIA's broader fascination with psychedelic drugs will be examined in a later chapter.

In 1955, Wasson ingested psychedelic mushrooms under the guidance of the Mazatec curandera María Sabina during a nocturnal *velada* in the Sierra Mazateca. He later described a cascading sequence of visions, each scene unfolding seamlessly from the last, as if palaces of colonnades, architraves, and courtyards of regal splendor were being built before his eyes in brilliant colors and precious materials, arranged with uncanny harmony and ingenuity.[131]

Wasson and CIA Involvement

Wasson would later claim that "out of the blue" a chemist named James Moore contacted him, expressed interest in joining his next expedition, and suggested that funding might be available from the Geschickter Fund for Medical Research. Subsequent document releases indicate that Moore was working with the CIA and that the Geschickter Fund was used as a conduit for MKUltra money; within CIA paperwork, Wasson's 1956 mushroom expedition appears as Subproject 58. Some researchers argue that Wasson must have been aware of Moore's institutional affiliations and of the fund's true role, though the degree of his knowledge and intent remains disputed.

Joining the expedition was the French mycologist Roger Heim, director of the National Museum of Natural History in Paris, who cultivated mushrooms from spores collected in Mexico. Heim sent preserved material to Albert Hofmann, the Swiss chemist celebrated for his work on LSD, and in 1958 Hofmann and colleagues succeeded in isolating the primary active compound, which he named psilocybin. [132]

The Sensational Impact of Wasson's Work

Wasson's adventures became widely known through his article "Seeking the Magic Mushroom," published in the May 13, 1957, issue of *Life* magazine under Henry Luce. Spanning many pages and illustrated with striking colour photographs and Heim's mushroom watercolors, the article detailed Wasson's psychedelic experiences and popularized the phrase "magic mushroom" in the English-speaking world. It caused a sensation, drawing waves of seekers, beatniks, and later hippies to the mountains of Oaxaca in search of their own visionary encounters.

While Wasson's reputation soared, the consequences for the Indigenous community he had visited were devastating, as later accounts from Huautla de Jiménez attest. [133] María Sabina was blamed for "revealing" the mushrooms to outsiders; she was ostracized, her house was burned, and her son was murdered, and she spent a period in jail before dying in poverty in 1985 at the age of 91.

Taken together, these histories show that psychedelic agents such as psilocybin have been woven into ritual, religious, and ceremonial life for millennia. Long before they were framed as "drugs," they were lauded as sacraments that could enhance spiritual experience and, in some traditions, may have helped shape the imagery and narratives that underlie contemporary religious beliefs.

LSD and Other Psychedelics

Psychedelics, a subclass of hallucinogens, are psychoactive substances that alter perception, mood, and a range of cognitive processes. Found naturally in plants and fungi or synthesized in laboratories, most "classic" psychedelics exert their primary effects by stimulating serotonin 5-HT2A receptors in the brain, with downstream influences on other neurotransmitter systems.[134] After ingestion or injection, effects usually begin within 20–90 minutes and can last from under an hour (as with smoked or injected DMT) to 8–12 hours (as with LSD), depending on the substance and route of administration. Although they are not considered physically addictive, repeated use leads to rapid tolerance and cross-tolerance, so that exposure to one psychedelic temporarily reduces sensitivity to others.

Psychedelics profoundly affect the senses, emotions, and perception of time. They may induce complex visual phenomena and alterations in thinking, or they may primarily intensify and distort ordinary sensory input, making colors seem more vivid and music more emotionally charged. Some users report synesthesia, such as "seeing" sounds as colors or patterns. At the same time, these substances can provoke acute anxiety, paranoia, or psychotic-like experiences, especially in vulnerable individuals or unsupportive settings. In a small minority, long-term disturbances such as persistent visual phenomena or flashback-like episodes have been described under the diagnosis of Hallucinogen Persisting Perception Disorder (HPPD), though this appears to be rare and remains poorly understood. [135]

DMT, José Capriles, and Ayahuasca

N,N-dimethyltryptamine (DMT) is particularly intriguing among naturally occurring psychedelics, not least because it is both a powerful hallucinogen and a compound that appears, in trace amounts, in mammals. In 2019, archaeologist José Capriles and colleagues analysed a 1,000-year-old ritual bundle from a rock shelter in the Bolivian Andes and identified residues of several psychoactive substances, including DMT and harmine, alongside cocaine. [136] The presence of DMT together with harmine—a monoamine oxidase inhibitor found in certain vines—suggests that people in the region may already have been combining plants in ways analogous to the ayahuasca brews used today. [137]

Today, ayahuasca—typically prepared from the leaves of *Psychotria viridis* (which contain DMT) and the vine *Banisteriopsis caapi* (which supplies MAO-inhibiting β-carbolines such as harmine)—is central to Brazilian syncretic religions and to a global circuit of retreats that claim to promote healing and spiritual insight. These contemporary uses sit in a complex relationship with Indigenous Amazonian traditions, raising recurring questions about cultural appropriation, commercialization, and psychological risk.

DMT has been detected endogenously in several animal species, and some researchers have argued that it may play a signaling role in the mammalian brain, though its physiological function, if any, remains uncertain. Psychiatrist Rick Strassman popularized the hypothesis that the human pineal gland might occasionally release surges of DMT, perhaps during REM sleep, near-death states, or extreme stress, and he has suggested that this could help explain some reports of vivid dreams, near-death experiences, and encounters with "entities" or aliens.[138] These ideas have attracted wide public interest but remain speculative; at present there is no direct evidence that naturally occurring DMT in humans routinely reaches levels high enough to produce full psychedelic experiences.

The Intriguing History of LSD

The history of LSD is equally captivating. In 1938, Swiss chemist Albert Hofmann, working at Sandoz Laboratories in Basel, first synthesized lysergic acid diethylamide (LSD-25) from lysergic acid, a derivative of ergot, the rye fungus. He was searching for compounds that might stimulate circulation and respiration without the dangerous uterine-contracting effects of other ergot derivatives.[139] LSD-25 showed no obvious clinical promise in initial animal tests and was set aside for several years.

In April 1943, Hofmann resynthesized LSD-25 and inadvertently absorbed a small amount through his skin during the preparation, experiencing unusual visual disturbances and restlessness. Intrigued, he deliberately ingested 0.25 mg (250 micrograms) of the substance three days later, producing what he later called the world's first intentional LSD trip.[140] In his memoir he recalled:

Kaleidoscopic, fantastic images surged in on me, alternating, variegated, opening and then closing themselves in circles and spirals, exploding in colored fountains, rearranging and hybridizing themselves in constant flux. Every sound generated a vividly changing image, with its own consistent form and color.

Hofmann became convinced that LSD had immense potential as a tool for psychiatry and for probing the biological basis of consciousness, but later lamented that its uncontrolled popularization—symbolized for him by Timothy Leary and the 1960s counterculture—derailed serious research for decades.

From 1947 onward, Sandoz made LSD available to psychiatrists and researchers under the trade name Delysid, supplying it at low or no cost and requesting access to clinical observations in return. In the 1950s, as interest in LSD spread, US government agencies also obtained the drug through official channels and diverted it into covert research programs, including the CIA's MKUltra project, which explored its potential for interrogation and psychological manipulation. Journalistic and historical accounts, such as Martin Lee and Bruce Shlain's *Acid Dreams*, [141] describe how US officials sought large stocks of LSD and even turned to Eli Lilly to develop independent synthetic production so they would not be dependent on Sandoz; claims that they aimed to procure enough for "100 million doses" are drawn from these secondary sources and should be treated as indicative of the scale of their ambition rather than as a verified figure. Sandoz did not enforce exclusive patent rights for LSD indefinitely, and by the early 1960s the compound could be manufactured by other laboratories, contributing to its diffusion beyond controlled medical settings.

The Unholy Alliance Between the CIA and Media Companies

An informal alliance between the CIA and major American media outlets took shape during Allen Dulles's tenure as CIA director. In his 1977 *Rolling Stone* article "The CIA and the Media," Carl Bernstein reported that, over roughly 25 years, more than 400 American journalists had undertaken secret assignments for the Agency, often with the knowledge or cooperation of their employers. [142] Dulles and his successors cultivated particularly close relationships with prestigious institutions, with CIA officials later describing The New York Times, CBS, and Henry Luce's Time Inc. as among their most valuable media partners.

Henry and Clare Boothe Luce: Advocates of LSD

From the late 1950s, Clare Boothe Luce—playwright, conservative icon, and wife of Time Inc. co-founder Henry Luce—began experimenting with LSD, initially under the guidance of writer Gerald Heard and, later, in sessions overseen by Dr Sidney Cohen, a Los Angeles psychiatrist involved in early government-backed LSD research. [143] Henry Luce had his first documented LSD session around 1960, also facilitated by Cohen, and quickly became an enthusiastic advocate in private. According to later accounts, he astonished a New York banquet audience in the mid-1960s by publicly acknowledging that both he and Clare had taken LSD—an extraordinary admission for one of America's most powerful media magnates. [144]

Time Inc.'s magazines, including *Time* and *Life*, went on to give LSD extensive, often breathless coverage in the early 1960s, helping to introduce the drug to a mass readership. Reflecting on this in his 1980 memoir, radical activist Abbie Hoffman remarked:

> I've always maintained that Henry Luce did much more to popularize acid than Timothy Leary. Years later I met Clare Boothe Luce at the Republican convention in Miami. She did not disagree with this opinion. America's version of the Dragon Lady caressed my arm, fluttered her eyes, and cooed, 'We wouldn't want everyone doing too much of a good thing.'" [145]

Legislative Attempts to Restrict LSD Use

Unaware of the CIA's role in helping to seed LSD into research and military networks, US legislators moved in the early 1960s to tighten control over the drug—steps that would sharply curtail legitimate psychopharmacological research. In 1962, the Kefauver–Harris Drug Amendments required that new medicines demonstrate both safety and efficacy through controlled clinical trials before they could be marketed, and tightened FDA oversight of investigational drugs. LSD, which Sandoz had previously distributed relatively freely to researchers as an experimental agent, now came under stricter regulatory control and could be used only under approved research protocols.

As recreational use and sensational media coverage intensified, President Lyndon Johnson singled out LSD in speeches and messages to Congress, urging tougher penalties. In March 1968, his administration backed legislation to make simple possession of LSD a federal offence, with proposed penalties of up to a 1,000-dollar fine and one year in prison for a first offence, and to increase penalties for manufacture and distribution. [146]

By late 1968, amendments to the Food, Drug, and Cosmetic Act had substantially stiffened federal penalties for LSD and other "dangerous drugs," even as enforcement remained patchy and research ground to a halt. [147]

CIA's Continued Stockpile

Despite tightening civilian controls, US military and intelligence agencies continued to hold and experiment with large quantities of LSD. At Edgewood Arsenal in Maryland, Army psychiatrist Lt. Colonel James Ketchum oversaw a series of controversial chemical-warfare experiments in which thousands of servicemen were exposed to LSD and other psychoactive agents.

In his memoir *Chemical Warfare: Secrets Almost Forgotten*, Ketchum recounts walking into his office one day in the 1960s to find a large black steel barrel filled with glass canisters of pure LSD—by his estimate enough to intoxicate several hundred million people and worth nearly one billion US dollars at then-current street prices (equivalent to several billion today). Within a week, the barrel had disappeared, and Ketchum wrote that he was never told where the drug went or what it was meant for, a small but telling vignette of the secrecy surrounding the US chemical-warfare and mind-control programs of the era. [148]

Modern Reflections and Ethical Considerations

The journey from Stone Age entheogenic rituals to contemporary psychedelic research reveals both human ingenuity and human vulnerability. As researchers once again explore the therapeutic potential of psychedelics for conditions such as depression, PTSD, and addiction, it is vital to acknowledge their dual nature: these substances can catalyze profound, positive transformations, yet they can also be misused in coercive or exploitative settings. Psychedelics thus sit uneasily at the boundary between healing tool and instrument of control.

Advances in digital technologies and neuropharmacology—including early efforts to use computational methods and artificial intelligence to design novel psychoactive compounds—underscore the need for robust ethical frameworks. The central challenge is to balance the potential benefits of these substances, whether for healing or for carefully framed spiritual exploration, with effective safeguards against manipulation, coercion, and subtle forms of psychological domination.

Conclusion

This chapter has traced the evolution of mind-altering substances from their enigmatic roles in ancient rituals to their modern deployment as tools for both healing and control. The historical and contemporary evidence suggests that while such substances have often been treated as gateways to the divine or to deeper self-knowledge, they can also be harnessed as mechanisms of manipulation. Recognizing this dual legacy is crucial as we confront the ethical complexities posed by modern neurotechnology and pharmacology.

The chapters that follow will examine how chemical, surgical, and digital methods of influencing the mind build upon these ancient foundations, raising difficult questions about autonomy, identity, and the future of human freedom.

CHAPTER 11 – THE US SEARCH FOR THE "TRUTH SERUM"

Introduction

In the chaotic aftermath of World War II, the United States faced unprecedented geopolitical and psychological challenges. Fearing that adversaries might develop methods to extract secrets from prisoners or turn ordinary citizens into unwitting assassins, US agencies launched a series of covert research programs. The quest for a "truth serum" was framed as a way to improve interrogation techniques, but it opened a Pandora's box of ethical and human-rights concerns that continue to reverberate today.

The Postwar Imperative: National Security Meets Psychological Science

In December 1948, Cardinal József Mindszenty, a prominent Hungarian archbishop and fierce critic of both fascism and communism, was arrested and accused of conspiracy, treason, and other fabricated offences against the new People's Republic of Hungary. Over 39 days he was deprived of sleep, starved, drugged, and beaten until he collapsed. At a five-day show trial in February 1949, he confessed in a flat, robotic manner to absurd allegations, including plotting a third world war for personal political gain, and was sentenced to life imprisonment.

The speed, theatricality, and apparent completeness of Mindszenty's confession alarmed US officials, who suspected that he had been subjected to some new form of "brainwashing" or mind control. This and other Cold War episodes spurred a broad governmental effort to develop techniques for extracting information from captives, "inoculating" US military and intelligence personnel against enemy interrogation, and exploring ways to manipulate or break the will of dissidents and foreign leaders. A reliable "truth serum" was especially coveted, but the ultimate fantasy was a method to compel individuals to perform acts—including assassinations—against their own interests and values.

To pursue this agenda, the US government established a clandestine unit at Fort Detrick, Maryland, tasked with devising chemical and biological tools for interrogation and behavioural control. Early work was organized under Project BLUEBIRD, which soon evolved into Project ARTICHOKE and, by the early 1950s, into the notorious MKUltra program. Many details remain obscured. In February 1973, shortly before leaving the Agency under pressure from President Richard Nixon, CIA Director Richard Helms ordered most MKUltra files destroyed, ensuring that only fragments of documentation and financial records would survive.

The existence of MKUltra and its precursors remained largely hidden until December 1974, when Pulitzer Prize-winning journalist Seymour Hersh published an exposé in *The New York Times* titled "Huge CIA Operation Reported in US Against Antiwar Forces, Other Dissidents." Based on information from inside sources, the article alleged that the CIA had compiled intelligence files on thousands of American citizens, conducted experiments on unwitting US civilians, targeted foreign governments, and plotted assassinations of foreign leaders. In response, President Gerald Ford established the United States President's Commission on CIA Activities within the United States (the Rockefeller Commission). Its 1975 report, together with subsequent Senate investigations led by the Church and Kennedy committees in 1975–1977, confirmed and expanded on many of Hersh's claims. [149]

Financial records and surviving memoranda showed that the CIA used a variety of methods to conceal its involvement in mind-control and behaviour-modification projects. Funds were channeled to universities, hospitals, and private researchers through apparently respectable foundations and cut-out organizations, through entities such as the Society for the Investigation of Human Ecology, through prominent intermediaries, and via law firms representing unnamed clients. In all, at least 149 MKUltra subprojects involved around 80 non-government institutions and some 185 outside researchers and assistants, many of whom were not fully informed about the Agency's role or ultimate objectives. [150]

Project Bluebird

Launched on 18 January 1950, Project BLUEBIRD was an early CIA effort to develop and systematize special interrogation methods using tools such as polygraphs, drugs, and hypnosis. Its founding memorandum described the goal as:

The immediate establishment of interrogation teams for operational support ... utilizing polygraph, drugs, and hypnotism to attain the greatest results in interrogation techniques. ... This interest stems from recent spy trials in Hungary and other satellite countries ...

BLUEBIRD methods were tested on prisoners at overseas interrogation centers, including facilities such as Camp King and Villa Schuster in occupied Germany and sites in the Panama Canal Zone. Detainees at these locations were subjected to harsh interrogation, including drugging, extreme deprivation, and other practices that would now be recognised as torture. Some accounts allege that prisoners died in the course of such experiments, although the surviving documentary record is incomplete and many details remain contested.

Project Artichoke – Broadening the Scope

In August 1951, Allen Dulles, then Deputy Director for Plans at the CIA, authorized the expansion of BLUEBIRD into Project ARTICHOKE. That September, research chemist Sidney Gottlieb was appointed chief of the Agency's Chemical Division within the Technical Services Staff and given responsibility for overseeing its activities. A CIA summary prepared in January 1975 described ARTICHOKE as:

> The Agency cryptonym for the study and/or use of 'special' interrogation methods and techniques. These include drugs and chemicals, hypnosis, and 'total isolation,' a form of psychological harassment ...

An internal memo from 1952 posed the chilling question:

> Can we get control of an individual to the point where he will do our bidding against his will and even against fundamental laws of nature, such as self-preservation?

This ambition closely anticipates the fictional "Manchurian Candidate" popularized by Richard Condon's 1959 novel—a programmed assassin triggered by a playing card—and illustrates how seriously Agency planners flirted with such ideas. Other ARTICHOKE documents describe experiments in which subjects were forcibly addicted to opiates, then pressured during withdrawal in an effort to make them more compliant.

One particularly disturbing memorandum considered how to deal with "unwilling subjects or subjects who cannot be trusted." While overt assassination was explicitly ruled out as incompatible with "our standards," options canvassed included irreversible psychiatric damage.

Lobotomy was discussed, apparently on the mistaken belief that it would reliably induce amnesia or fragment memory, but was ultimately rejected as "inhumane." The very fact that such proposals were entertained on paper underscores how far the program's planners were prepared to go in theory, even if not all options were implemented in practice. [151]

The Role of Key Figures

Journalist Gordon Thomas, in his book *Secrets & Lies: A History of CIA Mind Control & Germ Warfare*, describes a meeting in the early 1950s at Allen Dulles's Georgetown home attended by Canadian psychiatrist Ewen Cameron, British psychiatrist

William Sargant, and CIA officers Sidney Gottlieb and James Monroe

According to Thomas's account, the group discussed emerging mind-control techniques. Sargant emphasized that prolonged fasting, pain, and physical discomfort, if properly orchestrated, could induce anxiety, guilt, and nervous exhaustion—conditions he believed were fertile ground for reshaping belief systems. Monroe reportedly outlined the psychological effects of sensory isolation, while Cameron described his concept of inducing "traumatic psychological infantilism" through massive regression techniques. Convinced that these lines of work could crack the "mystery of brainwashing," Dulles agreed to commit substantial funding.

In this telling, Monroe was tasked with establishing the Society for the Investigation of Human Ecology as a front to channel research grants to universities and clinics, Sargant was to maintain liaison with Britain's Porton Down chemical-warfare establishment, and Gottlieb retained overarching authority for covert "executive action" capabilities, including targeted assassinations euphemistically referred to inside the Agency as "Executive Action."

While some elements of this narrative are supported by subsequent funding patterns and declassified records, Thomas's reconstruction of the meeting itself should be read as a synthesis of interviews and secondary sources rather than as a verbatim transcript. [152]

Project MKUltra – The Apex of Covert Mind-Control Research

On 13 April 1953, Project ARTICHOKE was superseded by the more expansive Project MKUltra ("MK" denoting projects run by the CIA's Technical Services Staff, and "Ultra" echoing the World War II codeword for highly classified intelligence).[153] Sidney Gottlieb was placed in charge.

MKUltra was closely linked to the Army's Special Operations Division at Fort Detrick, Maryland, headed by bacteriologist Frank Olson; this partnership allowed the CIA to draw on military expertise in chemical and biological agents while pursuing its own clandestine interests.

Funding and Expansion

MKUltra's research budget grew rapidly. In its first year, 1953, it reportedly reached around 1.04 million US dollars (approximately 12 million in today's terms), an extraordinary sum for a black program at the time. Roughly half of this was directed to pharmaceutical company Eli Lilly for the US production of LSD, reducing reliance on Sandoz.

The remainder was distributed across 18 separate projects, 15 of them designated "Top Secret." These funds flowed not only to internal CIA units and other federal agencies, such as the Bureau of Narcotics, but also to research groups at universities including Princeton, Stanford, and the University of Minnesota.

Donald Hebb

Donald O. Hebb, professor of psychology at McGill University in Montreal and later ranked among the most cited psychologists of the twentieth century, conducted pioneering work on sensory deprivation that would prove highly influential for MKUltra planners. In experiments at McGill, Hebb confined volunteers in small cubicles for up to six days, minimizing sensory input with translucent goggles, constant white noise via earphones, and cardboard cuffs that limited tactile stimulation.

A classified report from the early 1950s described striking effects. Within hours, many participants found it increasingly difficult to concentrate or think coherently; after a day or two, most reported vivid visual or auditory hallucinations, and the majority abandoned the study after two or three days. One subject experienced an enduring psychiatric breakdown. Hebb later acknowledged that some participants described the procedure as a form of torture. His findings were subsequently reflected in the CIA's 1963 Kubark Counterintelligence Interrogation manual, which recommended carefully managed isolation and sensory manipulation as means of disorienting and breaking down detainee.

Harris Isbell

Harris Isbell, a pharmacologist and director of the US Public Health Service's Addiction Research Center in Lexington, Kentucky, was contracted through MKUltra to study whether psychedelic substances such as LSD, mescaline, and DMT could increase susceptibility to hypnosis, induce delirium or psychosis, or otherwise weaken resistance.

His research involved hundreds of incarcerated men with histories of drug dependence, who "consented" to participate in exchange for access to their preferred substances—an arrangement that raises serious questions about coercion and voluntariness. Experiments were conducted on a dedicated ward under tight clinical control, but some protocols were extreme and hazardous. In one notorious trial, a small group of participants received escalating doses of LSD daily for roughly 80 consecutive days, reaching doses far above those required for a moderate psychedelic experience. From these and related studies, Isbell documented LSD's rapid tolerance and cross-tolerance with other psychedelics, noted that certain tranquilizers could blunt its acute effects, and concluded that the drug did not produce classic physical dependence—findings that would later shape both clinical and recreational use.

Louis Jolyon ("Jolly") West

Louis Jolyon West, an American psychiatrist who published widely on topics ranging from brainwashing to dissociation and schizophrenia, became one of the more controversial figures associated with MKUltra. He received a series of professional honours, including a Distinguished Professional Service Citation from the Oklahoma State Psychological Association, yet later journalists highlighted his proximity to covert research on hypnosis and psychedelic drugs.

Tom O'Neill and others have reported that, shortly after MKUltra was established, West wrote to Sherman Gifford (a pseudonym used by Gottlieb) proposing experiments that would combine hypnosis with psychoactive drugs in order to extract information from unwilling subjects, implant or alter memories, and embed complex messages in the minds of couriers.

West later served at Lackland Air Force Base, where he was involved in the "debriefing" and "deprogramming" of American prisoners of war returned from North Korea in 1953–54, many of whom had signed statements alleging that the US had engaged in biological warfare. Under intense questioning and the implicit threat of court-martial, virtually all recanted their confessions; the precise mix of psychological and pharmacological techniques used remains a matter of historical dispute.

In a 1956 proposal submitted to the CIA, West outlined methods for accelerating hypnotic induction, deepening trance states, and heightening suggestibility through sensory isolation and carefully structured environmental stress. He suggested that hypnotic suggestions could be used to induce marked psychological strain, confusion, or panic without leaving obvious physical traces—ideas that resonated strongly with MKUltra's aims.

West's MKUltra activities became public only after his death. But he did achieve an element of notoriety in 1962 when he injected LSD into an elephant in an Oklahoma City zoo, to determine the effect. The elephant dropped dead. The zoo was reimbursed secretly by the CIA.

Henry Beecher

Henry K. Beecher, professor of anesthesiology at Harvard Medical School, became widely known as a moral critic of research abuses after his 1966 New England Journal of Medicine article documenting 22 published studies in which subjects were exposed to significant risk without adequate informed consent.

Later archival work, however, revealed that Beecher himself had served as an adviser on classified human-experimentation projects during the 1950s. Documents from Harvard and government archives indicate that he participated in studies examining subjects' ability to keep secrets after consuming alcohol, amphetamines, barbiturates, and psychedelic agents, and that he visited CIA-linked facilities in Germany where harsh interrogation and high-risk drug experiments were conducted. In a 1952 report to the US Army Surgeon General, Beecher even speculated that LSD might have potential as a "non-lethal" incapacitating agent if introduced into the water supply of a target city—a suggestion more in tune with Cold War weapons thinking than with his later questionable reputation as an ethics reformer. Harvard still awards a prize in medical ethics, although it is no longer called the Beecher Prize.

Donald Ewen Cameron

Donald Ewen Cameron, a Scottish-born psychiatrist, held a series of prestigious posts: he chaired the Department of Psychiatry at McGill University, directed the Allan Memorial Institute in Montreal, served as president of the Canadian Psychiatric Association (1958–59), helped found the World Psychiatric Association (1961), and later led the Society of Biological Psychiatry. His legacy, however, is indelibly stained by his involvement in MKUltra-funded "psychic driving" experiments.

After taking up his McGill post in 1943, Cameron developed a 50-bed research and treatment unit at the Allan Memorial Institute. Under MKUltra Subproject 68, he designed an extreme regime that combined prolonged sensory deprivation, weeks of barbiturate-induced sleep ("narcosis"), repeated high-intensity electroconvulsive therapy far beyond standard clinical doses, and the administration of amphetamines and psychedelic drugs. Cameron believed that such procedures could "depattern" the mind—reducing patients to an infantile, highly malleable state— after which new behaviours and beliefs could be installed through repetitive taped messages ("psychic driving") and re-education. In reality, many patients suffered severe and often permanent cognitive and emotional damage; some lost basic life skills, personal memories, and the ability to live independently. Gottlieb later summarized one case in which:

> The shock treatment turned the then 19-year-old honours student into a woman who sucked her thumb, talked like a baby, demanded to be fed from a bottle, and urinated on the floor.

In the closing decades of the twentieth century, more than 250 former patients of the Allan Memorial Institute received compensation from the CIA and/or the Canadian government for injuries linked to Cameron's programs. In 2012, the student-run McGill Daily published the stark acknowledgement:

> To the patients of Dr. Ewen Cameron, our university was the site of months of seemingly unending torture disguised as medical experimentation.

MKUltra's Legacy

When John McCone succeeded Allen Dulles as CIA director in 1961, he ordered an internal review of the Agency's behavioural-research programs, including MKUltra. The Inspector General's 1963 report concluded that many of the projects raised serious ethical and legal concerns and acknowledged that some experiments had infringed the rights of US citizens. At the same time, it warned that full public disclosure would provoke intense criticism at home and abroad. The report recommended that the program be scaled back and its most controversial elements terminated. In an internal assessment, Gottlieb himself conceded:

> [I]t has become increasingly obvious over the last several years that the general area has less and less relevance to current clandestine operations. ... [These materials and techniques are] too unpredictable in their effect on individual human beings, under specific circumstances, to be operationally useful. ... Our operations officers ... seem to realize that, in addition to moral and ethical considerations, the extreme sensitivity and security constraints of such operations effectively rule them out.

MKUltra was officially halted in 1963–64, but portions of the work were re-labelled and continued under the umbrella of MKSEARCH, which focused on identifying, testing, and stockpiling chemical and biological agents that might have operational uses. Research into incapacitating chemicals, toxins, and delivery systems persisted into the early 1970s, often in conjunction with animal experimentation and military programs.

Gottlieb's Vietnam-Era Activities

The escalation of US involvement in Vietnam during the mid-1960s created new theatres for covert operations. Within this environment, the Phoenix Program, initiated in 1967–68, sought to dismantle the Viet Cong's political infrastructure through a combination of intelligence gathering, infiltration, capture, interrogation, and "neutralization." Investigations and survivor accounts describe a catalogue of abuses at Phoenix interrogation centers, including rape, severe beatings, the use of attack dogs, and electrical shocks delivered to sensitive areas such as the genitals and tongue, alongside other brutal techniques. Prisoners were often held without due process, and many did not survive, whether from execution, torture, or extrajudicial killing.

Several journalists and authors, including Gordon Thomas and Alex Constantine, have alleged that Gottlieb and CIA-linked technical units used the Vietnam conflict as a proving ground for more radical experiments. Thomas recounts two particularly disturbing episodes:

> - Electroconvulsive experiment: In one account, three psychiatrists were sent to Saigon in the mid-1960s, two of whom reportedly tested an adaptation of Ewen Cameron's "depatterning" techniques on prisoners. Detainees were subjected to rapid, repeated electroconvulsive shocks over several days in the hope of rendering them more pliable for interrogation. According to this narrative, the procedure produced no useful intelligence and resulted in the deaths of a number of prisoners.

- Brain-implant experiment: A second story describes a CIA-supported team, including a neurosurgeon, allegedly implanting electrodes into the brains of captured Viet Cong at Bien Hoa Hospital. The prisoners were then housed together and armed, while researchers attempted to trigger aggression or other behaviours remotely via radio-controlled stimulation of the electrodes. After several days without clear success, the experiment was abandoned, and the captives were reportedly executed and their bodies cremated.

Gottlieb retired from the CIA in 1972–73, in his mid-fifties, shortly after ordering the destruction of many MKUltra files at Richard Helms's direction. During congressional hearings in the 1970s, he gave guarded testimony, often claiming limited memory of specific projects and relying on the fragmentary surviving records. He died in 1999 at the age of 80.

Ethical Controversies and the Human Cost

Experimentation Without Consent

One of the most troubling features of BLUEBIRD, ARTICHOKE, and MKUltra was the routine use of human beings as experimental material without meaningful consent. Many "participants"—prisoners, psychiatric patients, soldiers, and other vulnerable people—were in no position to give informed, voluntary permission, and in some cases were not even told they were part of an experiment.

Exposed to invasive procedures that carried significant risk, and often caused lasting harm, they bore the human cost of programs designed in secrecy for abstract notions of security and advantage. This legacy of suffering continues to pose uncomfortable questions about the limits of state power and the obligation to respect human dignity even under conditions of perceived national emergency.

The Moral Price of National Security

National security was repeatedly invoked to justify these efforts, but the search for a truth serum exposed a troubling willingness to sacrifice individual autonomy for the promise of control. In practice, the experiments blurred or erased basic ethical boundaries, subordinating the rights of specific people to speculative strategic gains. The moral compromises embedded in these programs have cast a long shadow over public debate, serving as a cautionary tale about the dangers of unchecked authority and the ease with which scientific expertise can be weaponized against personal freedom.

Pont-Saint-Esprit and the Alleged LSD Experiment

Pont-Saint-Esprit is a small town on the banks of the Rhône in southern France. On 16 August 1951, while delivering mail, postman Léon Armunier was suddenly overwhelmed by nausea and vivid hallucinations: he later recalled feeling as though he was shrinking, while flames and serpents coiled around his arms. He fell from his bicycle and was taken to a hospital in Avignon, where he was restrained in a straitjacket. At the same time, local doctors were inundated with patients reporting nausea, vomiting, fever, and strange sensory disturbances.

In the days that followed, townspeople began to exhibit increasingly alarming symptoms. Some suffered hallucinations and convulsions; others ran naked through the streets; a boy attempted to strangle a relative; and at least one man leapt from a second-story window believing he could fly. Estimates of those affected range from roughly 250 to 500 people, with between five and seven deaths reported.

On 15 September 1951, the *British Medical Journal* published an account in which the treating physicians suggested that the outbreak was most likely caused by bread contaminated with ergot, a mould that can produce convulsions, vascular problems, and mental disturbances. In subsequent decades, however, journalists and some researchers have proposed that lysergic acid diethylamide (LSD), derived from ergot, might have been involved—either through deliberate contamination of bread, water, or air, or via some other form of exposure. Direct proof is lacking, but several circumstantial points have been cited:

- Passport and travel records:

Some authors note that Frank Olson and other scientists from the US Army's Special Operations Division at Fort Detrick—a unit involved in biological and chemical warfare research—were in France around the time of the incident. This has fueled speculation that Pont-Saint-Esprit might have been used as a covert field test. The documentation does not, however, show exactly where they went or what they did on the ground.

- Sandoz's involvement:

Almost immediately, a team of biochemists from Sandoz, the Swiss company that had synthesized LSD, travelled to Pont-Saint-Esprit to assist French authorities. Among them was Albert Hofmann, who had first created LSD from ergot derivatives. Sandoz had already been supplying LSD to US agencies for experimental purposes and had raised, in internal discussions, the possibility of using it or related compounds as incapacitating agents, including via water or aerosol delivery. This overlap of actors—ergot, LSD, Sandoz, Fort Detrick scientists—supports suspicion but does not establish intent.

- 1970s commission records:

In 2009, a journalist reported finding two intriguing references in the archived material of the 1975 Rockefeller Commission, which investigated CIA activities in the United States. One note, on White House stationery, reportedly mentioned Fort Detrick, Pont-Saint-Esprit, Frank Olson, Army Colonel Vincent Ruwet of the Special Operations Division, and two French intelligence officers in the same context. Another document allegedly instructed that files concerning Frank Olson and Pont-Saint-Esprit be hand-delivered to David Belin, the Commission's executive director, with a directive to "see to it that these are buried." The original context and full contents of these references remain disputed.

Taken together, these strands of evidence keep the possibility of a clandestine experiment open in the historical imagination, but they fall short of proving that US agencies deliberately dosed Pont-Saint-Esprit with LSD. The ergot-contaminated bread explanation remains the conventional medical view, while the LSD hypothesis stands as a provocative, circumstantial alternative that illustrates how Cold War mind-control programs colour our reading of otherwise baffling events.

Reflections on Legacy and Modern Parallels

Lasting Impacts on Ethical Research

The eventual exposure of MKUltra and related projects helped catalyze major reforms in research ethics. Today, formal protections such as informed consent requirements, independent ethics committees or institutional review boards, and clearer rules for government-sponsored research are meant to prevent a recurrence of such abuses. These safeguards are imperfect and unevenly applied, but they reflect a hard-won recognition that scientific and security objectives do not override the rights of human subjects. As new forms of neurotechnology, biometrics, and "enhanced" interrogation tools emerge, the ethical lessons of these Cold War programs remain critically relevant.

Modern Echoes in Digital and Neurotechnology

The chemical and behavioural tools of the mid-twentieth century may now look crude, but the underlying ambition—to bypass individual will and enforce compliance—persists. Contemporary technologies, including pervasive digital surveillance, psychological profiling based on big data, and algorithmic manipulation of information feeds, raise parallel concerns. Just as the promise of a truth serum offered a fantasized shortcut to obedience, modern systems threaten to enable quieter, more continuous forms of influence and control. Precisely because these mechanisms can be subtle and hard to detect, they demand sustained vigilance, transparency, and robust ethical oversight.

Conclusion

The US search for a truth serum marks a dark chapter in the history of intelligence and biomedical research. From Project BLUEBIRD through ARTICHOKE to the expansive MKUltra network, the attempt to chemically and psychologically override human will was inseparable from ethical violations and profound human suffering. These programs highlight how easily appeals to security can erode the protections owed to individuals.

As we confront emerging capabilities in neurotechnology and digital manipulation, we are forced to revisit the same fundamental dilemmas in new guises. The legacy of the truth-serum projects stands as a stark warning: when the capacity to shape thoughts, perceptions, and choices is concentrated in the hands of the state or powerful institutions, the potential costs to human dignity and freedom are immeasurable.

In the chapters ahead, we will examine how subsequent innovations in mind control have built upon these early experiments, shifting from chemical and surgical approaches to digital and algorithmic methods of influence. This historical lens underscores the importance of vigilance as we navigate the ethical challenges of a rapidly advancing technological landscape.

CHAPTER 12 – THE UK SEARCH FOR THE "TRUTH SERUM"

Introduction

Following World War II, the United Kingdom faced similar geopolitical anxieties as the United States. Fears that adversaries might develop methods to break prisoners' will or extract state secrets prompted UK defense and intelligence agencies to explore potential techniques for enhancing interrogation. The concept of a chemical agent capable of reliably lowering psychological defenses was viewed as a promising "shortcut" for intelligence operations, despite the lack of a solid scientific foundation. This chapter examines the evolution of the UK's efforts—from experiments at Porton Down to subsequent programs—and considers the ethical, scientific, and societal implications of these endeavors.

Research Facilities

Porton Down, a government facility some 90 miles south-west of London, was initially established during World War I to study chemical warfare. It later became the epicenter of the United Kingdom's research into chemical agents. Over subsequent decades its remit widened from defensive work on gases and protective measures to include experiments with incapacitating substances and the deliberate manipulation of human cognition and behavior, including LSD trials on servicemen. Related experiments were also carried out at other sites linked to interrogation and defense research, and British scientists investigated how drugs such as LSD affected mood, obedience, and memory.

As in the United States, detailed documentation on mind-control-oriented work at Porton Down and associated facilities remains sparse, but declassified records show that by the early 1950s British and American security services were sharing information on chemical and psychological methods of behavior control. Further investigations revealed:

- In May 1963, a Cabinet Defense Committee meeting chaired by Prime Minister Harold Macmillan approved "an increase in research and development on lethal and incapacitating chemical agents and the means of their dissemination," along with limited production of at least one agent, confirming that offensive-capable incapacitant research had become an explicit policy priority. [154]

- In November 1994, the UK government told Parliament that 72 service volunteers took part in LSD trials between 1962 and 1968. [155]

- A 2016 Defence Science and Technology Laboratory/MOD press release stated that since 1916 "over 20,000 volunteers had taken part in studies at Porton Down." It also mentioned the 1953 sarin-related death of Aircraftsman Ronald Maddison. [156] It added that Porton Down now produced only very small quantities of chemical and biological agents to support the development of medical countermeasures and protective equipment. [157]

- A 2002 BMJ report confirmed that the NHS agreed to pay a total of £195,000 in damages and £400,000 towards legal costs to 43 former psychiatric patients treated with lysergide (LSD) between 1950 and 1970, mainly at Powick Hospital in Worcester. [158]

- Further notes indicate that Powick's program ultimately treated 683 patients with LSD in 13,785 sessions before being discontinued.

- A 2004 report on Porton Down and subsequent commentary recorded that in 2006 three former servicemen who had volunteered for "common cold" tests at Porton in 1953–54, but were instead given LSD, received compensation of £10,000 each for their participation. [159] The men stated they were told the experiments concerned a cure for the common cold and that they were offered an extra week's pay for taking part. [160]

Key Personnel

Ronald Sanderson

Ronald Arthur Sandison trained in psychiatry and in 1949 became Deputy Medical Superintendent of Powick Hospital near Worcester. In 1952 he visited Sandoz in Switzerland, met Albert Hofmann, and returned with 100 vials of Delysid (LSD), which he began administering to patients at Powick, first to severely ill or "stuck" cases and then as an adjunct to psychotherapy for neurotic patients. Sandison later recalled telling these patients that this was an experimental drug that might help their condition, that it was considered safe, but that it could produce emotional disturbance.

In 1954 Sanderson and colleagues published "The Therapeutic Value of Lysergic Acid Diethylamide in Mental Illness" in the *Journal of Mental Science.* They reported on 36 patients at Powick, of whom 23 completed treatment and 14 were rated "recovered," with an average of about ten LSD sessions each. The paper drew wide attention and contributed to international interest in LSD therapy, reinforced by Sandison's presentations abroad and a growing series of meetings and conferences on the drug in the mid-1950s.

Sandison argued that LSD should be used in dedicated units, and Powick went on to open what is widely described as the world's first purpose-built LSD treatment clinic, a small ward on the hospital grounds that could treat about five patients at a time. Reports note that the unit was government-funded and that Sandison believed the support came through the Department of Health, although later commentators have suggested intelligence and defense interests also backed the early LSD work.

From 1952 to 1964 Sanderson treated hundreds of patients with what he termed "psycholytic therapy." He then resigned from Powick Hospital, disillusioned by the drug's spread into recreational use, and affected by Sandoz's discontinuance of LSD's manufacture. [161]

Harry Cullumbine

Captain Harry Cullumbine oversaw Porton Down's nerve-agent testing program which involved some 1,700 service personnel and lead to the 1953 death of Aircraftsman Ronald Maddison.

In 1953, at the request of the Secret Intelligence Service, Cullumbine began assessing the use of LSD as a potential aid to interrogation.

In 1956 the Defense Research Policy Committee authorized research into LSD as an incapacitating agent for individuals, groups, and crowds, with field trials run under the codenames Operation Moneybags, Operation Recount, and Operation Small Change, punning on the pre-decimal currency of pounds, shillings, and pence.

During one 1964 Small Change exercise, troops given oral LSD before a mock internal-security sweep quickly became ineffective: they moved slowly, broke into uncontrollable laughter, disobeyed orders, dropped their rifles, one climbed a tree to feed birds, and another repeatedly attacked a tree with a spade. After reviewing these trials, the Chemical Defense Advisory Board concluded in 1968 that LSD was "not a serious threat" in enemy hands and that the notion that it could be used as an incapacitating agent was "more magical than scientific." [162]

William Sargant

William Walters Sargant was one of the most controversial British psychiatrists of the 20th century. Born in 1907, his family had deep religious roots; his maternal grandfather was a Methodist minister, and five uncles were preachers. His brothers included

Norman, the Bishop of Mysore in South India, and Thomas, a politician and law reformer.

Sargant's work in LSD research and his broader psychiatric practices were marked by controversy, which will be elaborated further in this chapter.

Early Career and Medical Setbacks

Sargant studied medicine at St. John's College, Cambridge, graduating in 1930, and became a Member of the Royal College of Physicians three years later. During 1932–1933, Sargant published articles in The Lancet and the British Medical Journal, theorizing that the neurological effects of Addison's anemia (now known as Pernicious Anemia) were caused by iron deficiency and treatable with "intensive iron therapy." His mentor, Wilfred Harris, praised this as "a most important advance in neurological therapeutics." Unfortunately, Sargant's treatment proved disastrous, as the condition was later identified as stemming from the body's inability to metabolize vitamin B12, not iron. Excessive iron administration risked organ failure, and his approach was condemned by the medical community. This failure may have triggered one of Sargant's depressive episodes, though he attributed his decline to pulmonary tuberculosis. In 1934, he resigned from his position at St. Mary's Hospital.

Transition to Psychiatry

In 1935, Sargant pivoted to psychiatry and joined the Maudsley Hospital in South London. He became convinced that psychiatric illnesses were rooted in biological factors and thus amenable to physical treatments. Over the following years, he experimented with amphetamines, insulin-induced comas, Cardiazol-induced convulsions, thyroid hormone injections, and suprarenal gland injections. Despite these efforts, none demonstrated lasting benefit, though some patients responded positively to electroconvulsive therapy (ECT).

US Fellowship and Lobotomy Advocacy

In 1938, Sargant undertook a Rockefeller Foundation Fellowship at Harvard Medical School in Boston. During a visit to Washington, he met Walter Freeman, a proponent of lobotomy. After interviewing three lobotomized patients, Sargant resolved to introduce the operation to Britain. However, his British colleagues considered some of his treatments dangerous and imposed restrictions on his practices.

Professional Clashes and Religious Investigations

In 1944, Sargant came to wider prominence as co-author, with Eliot Slater, of the textbook *An Introduction to Physical Methods of Treatment in Psychiatry*, published by E. & S. Livingstone.academic. [163]

Later that year, he experienced another debilitating illness. During his convalescence, Sargant, influenced by his religious upbringing, explored accounts of religious conversion and the techniques employed by figures like 18th-century cleric John Wesley and contemporary brainwashers.

Following World War II, Sargant clashed with the Maudsley Hospital's clinical director, who denied him bed rights. His brother, then UK Secretary of Justice, warned him about his growing negative reputation. In his memoirs, Sargant reflected on this critique with striking narcissism and impoverished insight:

> My brother was shocked at the description ... I was said to be cruel and irresponsible, even being prepared to

recommend dangerous operations on a patient's brain. ... I seemed to have become something of a menace who refused to listen to my seniors' advice and insisted on pushing ahead with ill-balanced views.

After unsuccessful applications for other positions in the UK, Sargant spent a year as a visiting professor at Duke University, North Carolina, where he attended religious revival meetings and snake-handling ministries.

St. Thomas' Hospital and Controversial Practices

Returning to London in 1948, Sargant became head of the Department of Psychological Medicine at St. Thomas' Hospital. He utilized the newest drugs and physical treatments, establishing a "sleep room" where patients underwent sleep therapy for 20 hours daily over seven to ten days. They were allowed bedside meals and short walks to the toilet, often receiving ECT during treatment. Sargant stipulated that consciousness must be lightened prior to ECT, with fits modified using anesthesia and muscle relaxants.

Evidence suggests that at least one and possibly five patients died from Sargant's sleep therapy treatments—a fact conspicuously absent from his published works.

In 1954, Sargant suffered another bout of tuberculosis and depression, taking nearly a year off work. During this time in Majorca, Spain, he authored *Battle for the Mind*, published in 1957. This book documented practices in religion, voodoo, and cults, describing John Wesley as "the greatest brainwasher of the last two hundred years." Sargant argued that brainwashing techniques involved regressing individuals to a suggestible state through ritual, charisma, or physical deprivation, allowing new beliefs to then be implanted.

In 1967, he detailed his life story in *The Unquiet Mind.* After retiring from St. Thomas' in 1972, Sargant continued sleep therapy and insulin coma therapy for chronic schizophrenic patients in private practice. His book *The Mind Possessed: A Physiology of Possession, Mysticism, and Faith Healing* was published in 1973.

Controversial Reputation

To many in the medical profession, Sargant was perceived as a messianic, narcissistic figure surrounded by sycophantic supporters. He proclaimed his treatments as revolutionary, deriding dissenters and lamenting his lack of professorial elevation despite not completing requisite research degrees. He persisted with controversial treatments long after their dangers became clear.

I encountered Sargant in the late 1960s when he visited the research unit in Newcastle-upon-Tyne where I was working. Wearing a cape and carrying a cane, he made a brief show of inquiring about our work before pronouncing that we were wasting our time. The encounter left me with a lasting impression of theatricality and disdain that closely matched his reputation among many contemporaries.

Sargant achieved fame through *Battle for the Mind*, enjoying public lectures, broadcasting, and international conferences. His private practice flourished, granting him access to elite social circles.

Secret Involvement with Intelligence Agencies

Author Gordon Thomas, also a BBC producer, became friends with Sargant while featuring him in television programs. Sargant revealed his involvement with the CIA and British intelligence services on the condition that this information be disclosed only after his death. When Sargant died in 1988, his clinical records from St. Thomas' were missing, likely removed by MI5. Thomas relied on CIA documents to verify Sargant's covert activities.

These records confirmed Sargant's collaboration with MI5 and MI6 starting from the Cold War. In January 1946, he was issued a US Alien Registration Receipt Card and by 1970 he had traveled to Washington 31 times, regularly attending CIA Director Allen Dulles' Christmas parties. He conducted experiments on military volunteers at Maresfield Military Base and Porton Down, visited secret research centers, condoned fatal CIA experiments on German nationals, and maintained friendships with Sidney Gottlieb and Frank Olson, who kept him informed about the clandestine activities at Fort Detrick.

Sargant's Double Life and Ethical Failures

Sargant, like Ewen Cameron and Jolyon West, led a double life. While publicly presenting himself as a compassionate psychiatrist pushing the boundaries of therapy for the sake of his patients, he secretly collaborated with MI5, MI6, and the CIA to conduct highly unethical and sometimes lethal experiments.

Sargant may also bear some responsibility for the death of Frank Olson. By June 1953, Olson had grown deeply disillusioned with the CIA's activities and considered resigning. He sought psychiatric support from Sargant, who then breached professional confidentiality in a report intended for the CIA, in which he described Olson as "deeply disturbed over what he had seen in CIA safe houses in Germany ... displaying symptoms of not wanting to keep secret what he had witnessed." In November 1953, Olson fell to his death from the tenth floor of a hotel. While officially ruled a suicide, evidence suggests he was murdered.[164]

Challenges and Ethical Shortfalls

Like their US counterparts, British experiments in chemical interrogation frequently pushed the boundaries of ethical research. Volunteers—often drawn from vulnerable populations such as psychiatric patients or military personnel—were not always fully informed of the risks involved. In several instances, adverse effects led to later legal action and financial compensation. Taken together, these ethical failures tarnish the scientific record and stand as enduring reminders of the human cost of covert programs.

Proponents framed the search for a "truth serum" or incapacitating agent as a necessary response to national-security threats. Yet the methods employed routinely compromised individual autonomy. Experimentation without valid informed consent, combined with the use of drugs capable of profoundly altering or erasing memory, laid bare a persistent tension between the pursuit of state power and the inviolability of personal freedom. The resulting scandals left a legacy of controversy and mistrust that continues to shape debates on research ethics, human rights, and national-security policy.

Legacy and Modern Parallels

Lessons from the Past

Although the UK's chemical interrogation experiments never achieved the notoriety of projects like MKUltra, their impact endures. The ease with which state power can overstep ethical boundaries in the name of security remains a cautionary tale.

Modern Echoes

As technologies such as digital surveillance and AI-driven manipulation offer new ways to influence behavior, the legacy of these early experiments highlights the need for transparency, robust ethical oversight, and legal safeguards. The parallels between chemical interrogation methods and modern tools of influence underscore the importance of vigilance in protecting individual autonomy.

Conclusion

The United Kingdom's pursuit of a chemical "truth serum" represents a dark chapter in the history of state-sponsored research—one marked by scientific ambition and ethical compromise. From Porton Down's laboratories to subsequent experimental programs, UK researchers explored chemical agents' potential to control the human mind, often at great cost to individual dignity and autonomy.

Reflecting on these efforts, we are reminded that the drive for national security must be balanced against the protection of human rights. As new methods of manipulation emerge in the digital and neurotechnological realms, the ethical challenges posed by these early experiments remain as relevant as ever. In the chapters ahead, we will examine how chemical methods of mind control laid the foundation for surgical, digital, and algorithmic techniques—each presenting its own ethical dilemmas and implications for the future of human freedom.

The untimely death of Aircraftsman Ronald George Maddison

Maddison, a 20-year-old RAF mechanic, became a symbol of the risks of human experimentation at Porton Down. He volunteered for sarin tests in May 1953 in exchange for 15 shillings and a short leave pass, joining other servicemen in a gas chamber while wearing a respirator to prevent inhalation. Liquid sarin was applied to a cloth patch taped to his forearm to mimic battlefield exposure through contaminated clothing, but within minutes he developed classic signs of nerve-agent poisoning and quickly collapsed despite emergency treatment.

Maddison died later that day, and a secret inquest initially recorded a verdict of "misadventure" from a non-therapeutic experiment with a nerve agent. Decades later, a fresh inquest overturned this finding and concluded that he had been "unlawfully killed," transforming his case into a touchstone for ethical and legal scrutiny of Porton Down's volunteer program and the limits of state responsibility in dangerous research.

PART 3 - THE SURGICAL ASSAULT ON THE MIND

"What happens to free will when the mind becomes a battlefield for medical intervention?" –
Anonymous

CHAPTER 13 – EARLY PSYCHOSURGERY

Introduction

Throughout history, humanity has sought to control or modify behavior by altering the brain's structure. From the crude trephination practices of Neolithic societies to the invasive lobotomies and leucotomies of the mid-20th century, psychosurgery represents one of the most controversial attempts to address mental illness and, at times, enforce conformity. This chapter traces the evolution of surgical interventions aimed at modifying emotions and behaviors, highlights technical innovations and challenges, and considers the profound ethical dilemmas they raised.

Ancient Origins: Trephination and Early Interventions

Archaeological evidence suggests that as early as 7,000–10,000 years ago, some societies practiced trephination, the deliberate removal of a portion of the skull. Skulls from the Peruvian Andes, dating to roughly 2,000 years ago, show multiple trephine openings on a single cranium, and careful study of bone healing indicates that up to 80% survived the procedure. Although its precise purpose remains debated, historians and bioarcheologists generally argue that trephination was performed to relieve intracranial pressure after head trauma, treat neurological symptoms such as seizures, or expel malign spiritual forces thought to cause illness. [165]

The Advent of Psychosurgery in the Modern Era

António Egas Moniz

In 1949, Portuguese neurologist António Egas Moniz was awarded the Nobel Prize in Physiology or Medicine for his "discovery of the therapeutic value of leucotomy in certain psychoses." Eleven years earlier, working under Moniz's direction, neurosurgeon Almeida Lima had performed one of the first frontal lobe operations on a 63-year-old woman with severe anxiety, depression, paranoia, and auditory hallucinations. Using local anaesthesia, Lima drilled burr holes in the skull and injected alcohol into the frontal white matter to destroy neural connections; a 30-minute procedure that an independent observer judged to have eased her symptoms two months later. [166]

In December 1935 Lima refined the operation with a "leukotome," a narrow cannula tipped with a retractable wire loop that could cut cores of white matter and create more standardized lesions. Subsequent variants enlarged the skull openings to improve visual control and reduce operative risk.

A further major technical advance came in 1949, when Philadelphia neurologist Ernest Spiegel and neurosurgeon Henry Wycis introduced a stereotactic three-dimensional frame that allowed surgeons to target defined brain structures and create discrete lesions using electrical currents or freezing probes. [167] The stereotactic approach marked a shift from indiscriminate severing of frontal connections toward more localized interventions, and later developments—such as the implantation of permanent electrodes—permitted ongoing monitoring and increasingly precise lesion placement.

The Icepick Surgery Controversy

Walter Jackson Freeman II was one of the most polarizing figures in the history of psychosurgery. Intellectually gifted, he became professor of neurology at George Washington University School of Medicine in his early thirties, but colleagues and biographers alike have noted recurrent depressive episodes and hypomanic energy, along with an increasingly theatrical public persona. He cultivated a flamboyant image—goatee beard, broad-brimmed hat, cane, and carefully staged demonstrations—that made him instantly recognizable and helped turn lobotomy into a media spectacle rather than a strictly neurosurgical procedure.

Freeman first encountered António Egas Moniz's prefrontal leucotomy at the 1935 International Congress of Neurology and resolved to adapt the operation for use in the United States. In September 1936 he and neurosurgeon James Watts performed one of the first American frontal lobe operations on a 63-year-old woman with severe depression; her apparent improvement encouraged them to extend the technique to other forms of chronic mental illness. By 1942 they had reported a series of 136 patients with diagnoses including depression, obsessive–compulsive disorder, and schizophrenia, claiming "good" results in roughly two-thirds to three-quarters of cases, with a small but significant peri-operative mortality, and elaborated their views in their monograph *Psychosurgery: Intelligence, Emotion and Social Behaviour.*[168]

Determined to simplify and "democratize" leucotomy, Freeman developed the transorbital lobotomy in 1946. After rendering the patients unconscious with electroconvulsive therapy, he drove an ice-pick-like instrument through the upper eyelid and orbital roof into the frontal lobes and swept it laterally to sever white-matter connections, a technique that dispensed with the need for an operating theatre or neurosurgeon. Watts refused to participate in this method, regarding it as unsafe and unethical, and their professional partnership ended.

Backed by favorable press coverage, Freeman spent subsequent years touring state hospitals across the United States, demonstrating transorbital lobotomy on large numbers of patients in rapid succession. In the summer of 1951, for example, he performed scores of operations at multiple institutions over a few weeks; at one hospital in Cherokee, Iowa, three patients reportedly died among a group of 25, including two from intracranial haemorrhage and one from misplacement of the instrument into the midbrain. By the end of his career he had performed or supervised thousands of lobotomies, many of them transorbital, with a non-trivial operative mortality and a much larger number of patients left with profound cognitive and personality changes.[169]

While some contemporaries defended Freeman as a pioneer responding to the desperate conditions of overcrowded asylums before the psychopharmacological era, others saw his showmanship, the scale of his practice, and his disregard for informed consent and long-term outcomes as deeply troubling. Whatever weight one assigns to his own psychiatric instability, the failure of colleagues, administrators, and regulators to restrain his activities has come to be viewed as a major ethical failure in mid-twentieth-century psychiatry.

Ethical Controversies and The Human Cost

Balancing Therapeutic Promise and Risk

The published literature suggests that early frontal lobe psychosurgical procedures often benefited patients suffering from intractable anxiety, depression, or obsessive-compulsive disorders. In the hands of most surgeons, the risk of death or side effects—such as personality changes or epilepsy—was relatively small. When confined to patients with severe, prolonged disorders that had not responded to conventional treatments and who were often highly suicidal, these surgeries were considered justifiable, especially when performed using stereotactic methods.

Informed Consent and Vulnerability

A critical ethical failure of early psychosurgery was the frequent lack of informed consent. Many patients—already vulnerable due to the severity of their mental illness—underwent these invasive procedures without a clear understanding of the potential risks or outcomes. This exploitation of vulnerable populations remains a dark chapter in psychiatric history and continues to inform modern debates about the ethics of experimental medical interventions.

Legacy of Psychosurgery

The legacy of early psychosurgery serves as a stark cautionary tale. While it contributed to our understanding of the brain's functional anatomy and laid the groundwork for innovations like deep brain stimulation and other neuromodulatory therapies, it also came at a significant human cost in the hands of less scrupulous practitioners. Today, the field of neuroethics enforces stringent standards, emphasizing informed consent, meticulous patient selection, and rigorous ethical oversight to prevent a recurrence of past abuses.

Lessons for Modern Neuroethics

The early history of psychosurgery highlights how easily the urge to "repair" disordered minds can eclipse respect for individual autonomy and dignity. As techniques such as minimally invasive neuromodulation, deep brain stimulation, and brain–computer interfaces develop, these lessons remain acutely relevant. Researchers and clinicians must marry technical ambition with humility, building in rigorous consent processes, independent oversight, and long term follow up to minimize unintended harm and avoid repeating past abuses.

Conclusion

From the crude trephination of ancient societies to the highly invasive lobotomies of the twentieth century, psychosurgical practice showcases both human ingenuity and the ethical hazards of intervening directly in the brain. Even though later stereotactic procedures offered relief to some carefully selected patients, the wider record is indelibly marked by inadequate consent, uneven outcomes, and the suffering of vulnerable people.

As contemporary neurointerventions push into areas such as brain stimulation, "brain pacemaking," and closed-loop interfaces, the legacy of early psychosurgery stands as a warning and a guide: meaningful progress in treating mental and neurological illness must never come at the expense of human dignity, autonomy, and robust ethical safeguards.

The Curious Case of Eva Perón

Eva María Duarte de Perón (Evita) married Colonel Juan Perón in 1945 and quickly became his most powerful political ally, using her charisma and stagecraft to mobilize Argentina's poor, secure women's suffrage, and build a vast welfare network through the Eva Perón Foundation.

As uterine cancer advanced in the early 1950s, her pain, exhaustion, and increasingly militant rhetoric intensified; she reportedly began quietly assembling an armed workers' militia, financed through her foundation and supplied with thousands of weapons.

On the basis of skull imaging, archival material, and eyewitness testimony, several neurosurgical historians now argue that Eva Perón likely underwent a secret prefrontal lobotomy in Buenos Aires in the final months of her life. It was probably performed by visiting American neurosurgeon James L. Poppen, ostensibly to relieve cancer-related pain but also to blunt her agitation and political militancy.

The operation appears to have been carried out without her informed consent in a makeshift operating room under heavy guard, and was followed by rapid decline and death on 26 July 1952, aged 33. For many commentators, Eva Perón's story encapsulates both the power of psychological influence in politics and the ultimate abuse of psychosurgery as a tool of state control.

CHAPTER 14 – BRAIN STIMULATION

Introduction

The idea that electrical stimulation of the brain can alter behavior and perception has intrigued scientists for decades. From early experiments with implanted electrodes in animals to refined neuropsychiatric techniques, brain stimulation has emerged as a promising tool for therapeutic intervention and exploring human consciousness. This chapter chronicles the history of brain stimulation research, examines seminal experiments, and reflects on the ethical challenges posed when technology interfaces directly with the human mind.

Early Experiments and the Promise of Control

Long before modern neuromodulation techniques, researchers explored how targeted electrical impulses could influence behavior. Early studies demonstrated that applying electrical currents to specific brain regions could evoke sensations, trigger emotions, or suppress behaviors, laying the conceptual groundwork for later developments such as deep brain stimulation (DBS) and related neuropsychiatric interventions. These foundational experiments helped establish the principle that focal brain circuits can be reversibly modulated, paving the way for modern DBS in disorders like Parkinson's disease and obsessive–compulsive disorder.

The Breakthrough of José Delgado

In 1963, a dramatic event in a bullring on a ranch near Córdoba, Spain, showcased the potential of electrical brain stimulation. Neuroscientist José Manuel Rodríguez Delgado stepped into the ring with a red cape in one hand and a radio-controlled transmitter in the other, facing a 200-kilogram Spanish fighting bull that had been implanted with a "stimoceiver" in its caudate nucleus. As the bull charged, Delgado activated the transmitter and the animal halted abruptly, illustrating that electrical stimulation could rapidly interrupt aggressive motor behavior through a wireless device. [170]

This demonstration, captured on film and widely reported, became one of the most famous episodes in behavioral neuroscience. The event was later described in the press as "the most spectacular demonstration ever performed of the deliberate modification of animal behavior through external control of the brain," amplifying public and scientific interest in neuromodulation. [171]

Delgado's human work, while less theatrical, was no less ambitious. He reported that electrically stimulating discrete regions—especially within the temporal lobes—could evoke sensations such as elation, relaxation, deep concentration, "odd feelings," and colored visions, underscoring the potential of electrical stimulation to modulate mood and perception.

Stimulation of the septal area, near the brain's midline, frequently produced euphoria, suggesting that reward circuits could be accessed and temporarily reconfigured by precise current delivery.[172]

With animals, Delgado refined implantation strategies and tested closed-loop paradigms. In one line of experiments, he implanted electrodes in the amygdala of a chimpanzee named Paddy so that unpleasant stimuli were triggered whenever abnormal spindle-like brain activity was detected; this feedback loop reduced spindling and yielded a calmer behavioral state that often persisted after stimulation ceased. From such findings he extrapolated the idea of a "cerebral pacemaker" for conditions such as epilepsy, tremors, and obsessive behaviors, an intuition that prefigured later therapeutic devices.

Delgado's work with monkey colonies further examined how brain stimulation could shift social hierarchies. In one study, a dominant male directed aggression toward subordinates but not toward his mate. A female monkey named Elsa learned to press a lever that activated the stimoceiver in this dominant male, thereby inhibiting his aggression. These experiments raised provocative questions about whether rank, dominance, and conflict could be technically modulated rather than treated as immutable social facts.[173]

By the late 1960s and early 1970s Delgado had developed several interrelated devices. His "stimoceiver" integrated wireless radio transmission with intracranial recording and stimulation, enabling remote activation of implanted electrodes while allowing animals or patients to move freely. He later miniaturized it to roughly coin-size and added a telemetric system, advancing the feasibility of long-term brain–machine interfaces. In parallel, he invented the "chemitrode"—a multi-lumen device that could deliver controlled drug doses to specific brain regions while also recording electrical activity—anticipating later microinfusion and local drug-delivery systems.

Delgado also helped pioneer early forms of non-invasive brain stimulation. In his later years, Delgado "continued inventing, building a wireless human transcranial magnetic stimulator prototype," although his work in this area remained at the prototype stage rather than becoming an established clinical technique. Alongside these, he contributed to the development of an early cardiac pacemaker, demonstrating his broad interest in implanted regulators of bodily and cerebral function.

In a 1970 New York Times-style interview, Delgado reflected on the broader implications of these technologies:

> The human race is at a critical evolutionary juncture. We are on the verge of mastering our own mental functions through advancements in genetics … and through understanding the brain's mechanisms behind our behaviors. The pivotal question is: what kind of humans do we wish to create? [174]

He went on to argue that urban planning and human psychology were both poorly designed, and that the consequences of each "are dire," underscoring his concern that society lacked the ethical framework to match the emerging power of neuromodulation.

Despite his technical originality and conceptual foresight, Delgado later expressed frustration that controversy and shifting funding priorities had prevented him from developing his ideas in depth and warned that scientific and technological advances would proceed regardless of ethical qualms.

His work, however, remains a touchstone for debates about the boundaries between therapeutic intervention, enhancement, and manipulation in neuropsychiatry.

The Controversial Robert Heath

Robert Galbraith Heath, a neurologist and psychiatrist, is remembered as one of contemporary psychiatry's most controversial figures. Driven by a conviction that biological brain mechanisms underlay major mental illnesses, he became obsessed with the idea that he might discover the cause of schizophrenia and win a Nobel Prize. In 1949 he accepted the founding chair of a combined Department of Psychiatry and Neurology at Tulane University in New Orleans, a post that gave him access to large numbers of patients for invasive research. At Tulane he assembled a multidisciplinary team and developed a program using chronically implanted depth electrodes to record from and stimulate deep brain structures in humans. [175]

Schizophrenia and the Taraxein Myth

At the 1956 annual meeting of the American Psychiatric Association, Heath announced that he had isolated a protein from the blood of patients with schizophrenia, which he called Taraxein, and that injections of this substance into non-psychotic volunteers produced what he described as the "full symptoms of schizophrenia." He reported similar effects in monkeys and claimed that Taraxein induced abnormal electrical activity in the septal region which he believed was characteristic of schizophrenia. The work attracted major media attention, and a Tulane press release even suggested that the finding could rank among the most significant advances in psychiatry, prompting speculation that Heath might be a Nobel contender.

However, no independent laboratory was able to replicate Heath's results, and subsequent biochemical analyses failed to confirm the existence of Taraxein as a specific, schizophrenia-linked protein or antibody. Later commentators have suggested that Heath's senior technician—who had criminal connections and was later murdered—surreptitiously administered a psychedelic drug to produce the dramatic effect.

Schizophrenia and the Epilepsy Myth

In parallel with his Taraxein work, Heath pursued the idea that schizophrenia might be a form of "deep epilepsy" arising from pathological activity in limbic structures that subserve memory and emotion. In the early 1950s he reported on depth-electrode studies in roughly 20 or more patients diagnosed with schizophrenia, claiming that recordings from the septal region showed spike-and-wave epileptiform discharges absent in other diagnostic groups. He further asserted that electrical stimulation of the septal area produced marked clinical improvement in a subset of patients.

Yet even in Heath's own series, severe adverse events occurred: several subjects developed seizures and other complications following electrode implantation and stimulation, and some died in temporal proximity to the procedures. Subsequent investigators using depth electrodes in patients with schizophrenia did not find the distinctive "septal epilepsy" pattern that Heath described, and modern epilepsy research has not validated his claims.[176] As a result, his proposed epileptic mechanism for schizophrenia is now regarded as an erroneous and potentially harmful line of inquiry.

Heath's Claim That He Could Change Sexual Orientation

Heath's reputation deteriorated further after his experiments on sexuality and brain stimulation became public. In 1972 he published accounts of a 24-year-old white male gay sex worker, codenamed B-19, who had a history of psychiatric admission and substance misuse. Heath maintained that B-18 sought treatment both for depression and his homosexuality, and hypothesized that his homosexuality was associated with temporal lobe epilepsy.

Electrodes were surgically implanted over the cortical surface and within deep brain structures, including the septum and thalamus. When B-19 was shown a 15-minute film depicting heterosexual intercourse, he reportedly expressed disgust and showed no EEG change suggestive of epileptiform activity. However, when the septal electrodes were stimulated, B-19 described intense sexual arousal, and when given a hand-held button he self-stimulated the septal region hundreds of times, reported "feelings of pleasure, alertness, and warmth," and protested vigorously when the session ended.

After repeated sessions of septal self-stimulation, B-19 viewed the same heterosexual film and while stimulation was applied he masturbated to orgasm, but again Heath reported no characteristic seizure activity on the EEG. When the patient later expressed a desire to attempt heterosexual intercourse, Heath and colleagues arranged for a 21-year-old female sex worker to spend time with him in a hospital laboratory room. According to Heath's published account, B-19 engaged in intercourse successfully, and depth recordings showed spike-and-wave activity in septal and thalamic regions around the time of orgasm. Heath then prescribed further self-stimulation and masturbation as a "treatment" protocol and claimed that B-19 had subsequently formed a close relationship with a married woman, while acknowledging that the patient also continued to have sex with men for money.

Later analyses, including interviews and re-examinations of Heath's publications, have concluded that his claim to have "cured" homosexuality was unsubstantiated and methodologically flawed, and that the experiments were ethically indefensible even by the standards of the time. Contemporary scholars emphasize that any transient heterosexual behavior under intense conditioning and stimulation does not equate to a change in enduring sexual orientation. [177]

The Mounting Criticism

By the 1970s, criticism of Heath's work—from both scientific and ethical perspectives—had intensified. Other researchers could not replicate either the supposed septal epileptiform abnormality in schizophrenia or the biochemical finding of

Taraxein in the plasma of patients. Heath's heavy reliance on invasive procedures in vulnerable populations, including psychiatric inpatients and prisoners, attracted growing concern as research ethics standards evolved.

In his 1973 book *Brain Control*, psychologist Eliot Valenstein reviewed Heath's stimulation studies and reported that many of Heath's patients, when interviewed independently, said that septal self-stimulation had not in fact produced the intense pleasure described in the published papers, but that they had continued the behavior because they believed it was expected of them.[178] Valenstein and other critics argued that Heath's conclusions frequently exceeded what his data could support and that his enthusiasm for technological "control" of behavior had led him to minimize risks and overstate benefits.

Heath's CIA Activities

Declassified documents from the US Central Intelligence Agency show that in the 1950s Heath's laboratory participated in MKUltra-related research exploring psychoactive drugs and brain function. According to investigative accounts and partial CIA records, he received funding—reported in some sources as a five-year grant totaling around $60,000—to study the behavior of subjects with implanted electrodes during administration of LSD-25 and mescaline. Separate reports indicate that Heath also tested bulbocapnine, a dopamine depleting alkaloid, on inmates at the Louisiana State Penitentiary at Angola, describing the induction of a state of catalepsy marked by rigidity and trance like immobility. Heath's willingness to use prisoners and psychiatric patients in high risk experiments aligned uncomfortably with this secretive agenda and has further darkened assessments of his legacy. [179]

Assessment

Taken together, Heath's career illustrates how scientific brilliance and technical innovation can be profoundly compromised by poor methodology and disregard for ethical constraints. He contributed to the early development of depth-electrode recording and brain stimulation in humans, but his unreplicable findings, grandiose claims, and exploitation of highly vulnerable subjects inflicted lasting damage on the reputation of neuropsychiatric research. Many historians of psychiatry now view him less as a visionary than as a cautionary exemplar of how not to conduct human brain experimentation.

Francis Ervin's Study of Rage and Violence

Francis Raymond Ervin was a relatively reserved figure who nonetheless became a central target of the anti-psychiatry and anti-psychosurgery movements of the 1970s. After earning his medical degree from Tulane University in New Orleans in 1951, he completed a three-year residency in psychiatry and neurology and was board-certified in both specialties before joining Robert Heath's research team at Tulane in the mid-1950s.

Ervin gained widespread notoriety in 1970 with the publication of *Violence and the Brain*, co-authored with neurosurgeon Vernon H. Mark. [180] Written for a general audience, the book argued that some forms of violent behavior might be rooted in identifiable brain pathology and explored the neurological substrates of rage and aggression. Drawing on animal and human data, Mark and Ervin emphasized the role of the amygdala and its connections with the hypothalamus in the production of anger and defensive rage. They cited, among other examples, the case of Charles Whitman, the "Texas Tower Sniper," who, in 1966, killed his mother and wife before fatally shooting 15 people and wounding 31 others at the University of Texas at Austin. An autopsy revealed a brain tumor impinging on his right temporal lobe near the amygdala, which they suggested might have contributed to his uncontrolled violence. Mark and Ervin proposed that if a subset of violent acts arose from such brain disease, early identification and targeted intervention might help prevent some episodes of extreme aggression.

In their clinical work, Mark and Ervin examined several groups: patients with temporal lobe epilepsy, individuals with severe impulse-control problems, and prisoners with histories of serious violence. They observed that many subjects in these groups showed a distinctive pattern of behavior that included:

- A history of recurrent physical assaults, often within the family

- Episodes of "pathological intoxication," in which relatively small amounts of alcohol precipitated disproportionate, senseless violence

- Impulsive or inappropriate sexual behavior, sometimes including sexual assaults

- Frequent traffic violations or serious automobile accidents.

They termed this constellation the "dyscontrol syndrome" and argued that it could point to an underlying neurological disorder in some chronically violent individuals.

Across both prison and non-prison cohorts, Mark and Ervin reported that roughly half of the subjects with dyscontrol syndrome further exhibited features suggesting an epileptic basis for their outbursts—such as stereotyped forewarning of an impending rage attack, altered consciousness during the episode, and post-attack drowsiness or lethargy. They estimated that the incidence of epilepsy in their prisoner group was about ten times higher than in population estimates of the time and noted that EEG recordings revealed abnormal brain waves in about half of the prison sample and roughly one-third of the non-prison group. On this basis, they concluded that structural or functional brain disease could underlie violent behavior in a subset—though certainly not all—of those with the dyscontrol syndrome.

In the half century since the publication of *Violence and the Brain,* epilepsy and neuropsychiatry research has broadly supported Mark and Ervin's claim that some patients with violent "dyscontrol" episodes have an underlying seizure disorder, particularly involving the amygdaloid nucleus, while also making clear that epilepsy explains only a small, clearly defined subset of such cases. [181]

Mark and Ervin provided detailed depth-electrode studies of four patients, one of the most cited being "Julia," a woman with temporal lobe epilepsy and severe episodic violence in whom amygdala stimulation could reliably provoke rage and threat displays. Her case is detailed in the textbox below.

The surge of anti-psychiatry and anti-psychosurgery activism in the early 1970s, particularly in response to proposals for "violence centers" and to media coverage of *Violence and the Brain*, led to intense scrutiny and criticism of Mark and Ervin's work, and Ervin increasingly shifted his focus from human psychosurgery to primate research. In 1972 he founded the Behavioural Science Foundation in St Kitts, establishing laboratories to study brain and behavior in the island's native vervet (cercopithecine) monkeys, a program that later drew protest from animal-welfare advocates concerned about the use of non-human primates in invasive research.

Ervin joined McGill University's Department of Psychiatry in 1979 and continued his work there on the neurobiology of behavioral control, ultimately retiring as Professor Emeritus in 2009. He maintained his scientific activities in St Kitts well into later life, co-founding the Ripple Institute with Roberta Palmour and continuing to oversee monkey research until his death at his home in St Kitts on 24 April 2015 at the age of 88. Over his career he authored nearly 300 scientific papers and at least one major book, leaving a substantial—if controversial—legacy in the study of brain mechanisms underlying violence and impulse control.

Julia's Case: Brain Stimulation and the Treatment of Rage

Julia's case, reported in 1969 when she was 21, illustrates the challenges of treating severe temporal lobe dysfunction and rage and is discussed in detail by Vernon Mark and Frank Ervin in *Violence and the Brain*. Her difficulties began early: at age two she developed mumps encephalitis, and by age ten epileptic seizures had appeared. At times these took the form of generalized convulsions, but most were temporal lobe seizures—brief lapses of consciousness with staring, lip smacking, and chewing movements. Often, after such episodes, a wave of panic would drive her to flee aimlessly. Between seizures she suffered severe temper outbursts, followed by intense remorse, and over the years she made four serious suicide attempts.

Her impulsive violence became alarming. On twelve separate occasions she seriously assaulted others apparently without provocation. In one episode, in a movie theatre, she stabbed a girl in the heart; in another she seized a pair of scissors and stabbed a nurse in the chest.

Despite prolonged treatment at three major North American centers—including multiple drug regimens and more than sixty electroconvulsive therapy sessions—her condition showed little improvement. Electroencephalography consistently revealed epileptic spike activity in both temporal lobes, and imaging demonstrated atrophy of the right temporal lobe. To better understand and potentially modulate her symptoms, electrodes were stereotactically implanted into both temporal lobes. Once she had recovered from surgery, depth EEG recordings localized abnormal activity in the amygdaloid nuclei bilaterally, and electrical stimulation of these regions reproduced symptoms resembling the onset of her habitual seizures and evoked episodes of intense anger with facial grimacing and threat behavior.

A lesion was first made in her left amygdaloid nucleus, but her seizures and attacks persisted, so an ablative lesion was later made in her right amygdala as well. By 1973, following these interventions, Julia's assaultive behavior had ceased. Although she continued to experience impulsive anger and psychotic symptoms, she no longer had rage attacks; she was able to live at home, attend adult-education classes daily, sing in a choir, and pass her high-school equivalency examinations.

Ethical Considerations

Manipulation Versus Therapy

The ability to alter behavior, mood, and even aspects of personality through direct brain stimulation raises profound ethical dilemmas. On one hand, neuromodulation offers potential relief for patients with severe, treatment-resistant conditions such as advanced Parkinson's disease, obsessive–compulsive disorder, or major depression. On the other hand, the same tools could be deployed to influence behavior without valid informed consent, especially in institutional settings, blurring the boundary between therapy and social control. The fine line between legitimate clinical intervention and coercive manipulation underscores the need for rigorous ethical oversight, independent review, and clear limits on non-therapeutic or punitive uses of neuromodulation. [182]

Informed Consent and Vulnerable Populations

Historically, some brain-stimulation studies recruited highly vulnerable subjects, including psychiatric inpatients and prisoners, whose capacity to refuse participation was often compromised. Contemporary research ethics insist that any invasive neurointervention must be based on robust informed consent, typically overseen by institutional review boards or research ethics committees. Yet consent in this context is complex: neuromodulation can alter cognition, mood, and decision-making, and many patients have impaired capacity or poor long-term recall of the information provided, which raises questions about how stable consent remains over time. These concerns are amplified for detainees and offenders, where offers of neurointerventions may be experienced as coercive "choices" tied to sentencing, confinement, or access to care. Ensuring that participants understand risks and benefits, can withdraw without penalty, and are protected against both overt and subtle coercion is central to respecting autonomy and preserving human dignity.

The Future of Neuroethics

As brain-stimulation methods become more sophisticated and increasingly integrated with digital systems and brain–computer interfaces, new ethical questions arise about autonomy, identity, and mental privacy. Neural devices that record and modulate brain activity generate sensitive data about thoughts, intentions, and emotional states, raising concerns over who owns this information, how it may be used, and how it can be protected from misuse—including "brainjacking" or unauthorized access. At the same time, post-stimulation changes in personality or sense of self have led ethicists to revisit notions of authenticity and personal identity, asking when a technologically altered self still counts as "one's own."

These developments highlight the importance of coupling scientific progress with robust ethical guidelines, legal safeguards, and ongoing public deliberation so that neuromodulation enhances human wellbeing without eroding fundamental rights and freedoms.

Conclusion

Brain stimulation has come a long way since its experimental beginnings. The pioneering work of figures like José Delgado demonstrated the potential for electrical impulses to modify behavior, paving the way for treatments that offer hope for neurological and psychiatric disorders once deemed untreatable. However, the power of brain stimulation also introduces significant ethical challenges.

As neurotechnologies continue to evolve, it is critical to maintain a vigilant ethical framework that prioritizes individual autonomy and dignity. Advances in neurotechnology must be harnessed to enhance — not compromise — human freedom. In the chapters ahead, we will explore how brain pacemaking and cognitive augmentation build on these early interventions and consider how they may shape society in both beneficial and potentially perilous ways.

Four Core Neuroethical Principles in Brain Stimulation

- **Autonomy**
  Respect the person's right to decide whether to undergo brain stimulation, based on clear, understandable information and freedom from coercion.

- **Beneficence**
  Aim to promote the patient's well-being, using brain stimulation only when there is a realistic prospect of meaningful clinical benefit.

- **Non-maleficence**
  Minimize risks and avoid causing harm, including surgical complications, unwanted psychological changes, and long-term device-related burdens.

- **Justice**
  Ensure fair access to beneficial interventions, avoid targeting vulnerable groups for risky experiments, and prevent discriminatory or punitive uses of neuromodulation.

CHAPTER 15 – BRAIN PACEMAKING

Introduction

Deep brain stimulation (DBS) has revolutionized the treatment of neurological and psychiatric disorders by modulating neural circuits directly. As the technology evolved, researchers envisioned adaptive systems that could respond in real time to the brain's changing needs, much like a cardiac pacemaker regulates heart rhythm. These closed-loop systems, often referred to as "brain pacemakers," hold immense promise for restoring or enhancing brain function. However, they also raise profound ethical and social questions concerning autonomy, identity, and the potential for misuse.

Helen Mayberg: Pioneer in Psychiatric Pacemaking

Helen Susan Mayberg, a neurologist and trailblazer in the use of deep brain stimulation (DBS) for psychiatric illness, serves as Professor of Neurology, Neurosurgery, Psychiatry, and Neuroscience at the Icahn School of Medicine at Mount Sinai. She is also the founding director of the Nash Family Center for Advanced Circuit Therapeutics at Mount Sinai, where her work focuses on circuit-based interventions for mood and related disorders.

Mayberg proposed that depression does not arise from a single brain region or neurotransmitter abnormality but from dysfunction within an integrated network of pathways, including limbic structures that regulate emotion, behavior, and memory. Within this framework, she identified the subcallosal cingulate (SCC) region as a key node for intervention, based on converging evidence:

- Structural and functional abnormalities in the SCC in depressed patients

- MRI findings of reduced regional volume

- Functional imaging showing increased SCC activity during sad mood and decreased activity following successful treatment

- Post-mortem studies reporting reduced glial cell density in this area.

In 2003, Mayberg and neurosurgeon Andres Lozano implanted DBS electrodes in the SCC of patients with treatment-resistant major depressive disorder. Quadripolar electrodes were positioned bilaterally, and intraoperative testing showed that stimulation at specific contacts could reliably induce calm, mood elevation, heightened awareness, and renewed interest in the environment. These effects were reproducible and absent during subthreshold or sham stimulation. The electrodes were then connected to implanted pulse generators in the chest to deliver chronic DBS.

To assess clinical outcomes, the team used standard psychiatric rating scales, including the 17-item Hamilton Depression Rating Scale (HDRS), on which a reduction of at least

50 percent indicates a treatment response and a score of 8 or less indicates remission.

In a 2005 *Neuron* report on the first six patients to receive SCC DBS, pretreatment HDRS scores were in the severe range (mid-20s to high-20s). After six months of continuous stimulation, three patients achieved remission or near-remission and a fourth showed marked improvement.[183] Responders described renewed engagement with family and social life, reduced apathy, better sleep, and a restored capacity for pleasure. Functional imaging demonstrated patterns of network change resembling those seen with effective antidepressant medication.

A placebo effect was further discounted in a case where stimulation was discontinued without the patient's knowledge: within four weeks she developed reduced initiative and energy, though not full depressive relapse. When stimulation was reinstated, again without her awareness, her prior level of functioning returned within a week.

Subsequent long-term studies have supported the therapeutic potential of DBS for both unipolar and bipolar depression, as well as for conditions such as obsessive-compulsive disorder, with symptom recurrence commonly observed when stimulation is interrupted or devices fail.[184]

Mayberg's impact on neuropsychiatry has been widely acknowledged. She has authored more than 200 peer-reviewed publications, secured major research funding over her career, and received numerous honors, including the Gold Medal Award from the Society of Biological Psychiatry (2014) and the Steven E. Hyman Award for Distinguished Service to the Field of Neuroethics (2018).

DBS in Other Psychiatric Disorders

DBS has shown promise in several other psychiatric conditions. In 2008, a study in *Molecular Psychiatry* reported outcomes in 26 patients with severe, treatment-resistant obsessive-compulsive disorder who received DBS targeting the ventral anterior limb of the internal capsule/ventral striatum; about two-thirds of patients were classified as responders, with marked reductions in OCD symptoms and improved functioning, and most adverse effects were transient or stimulation-related. In 2009, the US Food and Drug Administration granted a Humanitarian Device Exemption (HDE) for Medtronic's Reclaim DBS system for chronic, severe OCD, making it the first DBS device formally approved for a psychiatric indication. [185]

Ongoing research is investigating DBS as a potential intervention for other refractory conditions, including anorexia nervosa, Tourette syndrome, post-traumatic stress disorder, aggressive and impulsive disorders, substance use disorders, and cognitive and executive impairments following traumatic brain injury, with early trials suggesting possible benefits in carefully selected patients but underscoring the need for larger, controlled studies.

Emergence of Closed-Loop Systems

Traditional DBS systems deliver constant electrical stimulation to target brain regions, regardless of moment-to-moment neural fluctuations. While effective for many conditions, this open-loop approach does not account for the dynamic nature of brain activity. For example, patients with treatment-resistant depression may experience varying neural states throughout the day, yet an open-loop DBS device continues to deliver uniform stimulation

Around 2018, a team led by neurosurgeon Edward Chang at the University of California, San Francisco, began testing DARPA- and BRAIN Initiative-funded closed-loop deep brain stimulation systems—essentially "brain pacemakers" that sensed pathological activity patterns and delivered responsive stimulation. The system identified an excess of beta-frequency (13-30 Hz) waves in the connections between the amygdala and hippocampus as a biomarker for depression. The device delivered stimulation only when this abnormality was detected. The targeted area in their studies was the ventral capsule/ventral striatum (VC/VS).[186]

Side Effects of DBS

Reported side effects of DBS include anxiety, euphoria, disinhibition, aggression, altered energy levels, and changes in libido. Most of these effects are transient and may depend on the target site or stimulation parameters.[187] Permanent side effects are rare and have been primarily documented in DBS studies for movement disorders like Parkinson's disease, though these may be related to disease progression rather than DBS itself. A key advantage of DBS is its reversibility; the device can be removed if side effects become intolerable.

INDICATION	MAIN DBS TARGET(S)	TYPICAL RATIONALE / NOTES
Treatment-resistant unipolar & bipolar depression	Subcallosal cingulate (SCC/Cg25)	Modulate overactive limbic–frontal mood network; Mayberg/Lozano series and later long-term follow-up.
Severe, refractory OCD	Ventral anterior limb of internal capsule (VALIC), ventral striatum / nucleus accumbens	Disrupt pathological cortico-striato-thalamo-cortical loops; 26-patient *Molecular Psychiatry* cohort; basis for Medtronic Reclaim HDE.
Anorexia nervosa (severe, chronic)	Subcallosal cingulate, nucleus accumbens	Target mood/anxiety circuitry and reward–motivation pathways; small open-label trials only.
Tourette syndrome	Thalamus (centromedian–parafascicular complex), globus pallidus internus, anterior limb of internal capsule	Reduce tic frequency and severity via motor and limbic circuit modulation; multicenter case series and registries.
Aggression / impulsive violence	Posteromedial hypothalamus, amygdala, ventral striatum	Aim to dampen pathological aggression circuits; very small, highly selected case series.
Substance use disorders	Nucleus accumbens / ventral striatum	Interfere with craving and reward salience; early pilot studies, mostly in alcohol and opioid dependence.

INDICATION	MAIN DBS TARGET(S)	TYPICAL RATIONALE / NOTES
Cognitive / behavioral deficits post-TBI	Central thalamus, frontal/striatal targets	Enhance arousal, attention, and executive function; early proof-of-concept studies in chronic brain injury.

Prospects for Personal Enhancement

Reports from DBS studies in treatment-resistant depression show that higher-intensity stimulation can sometimes produce acute euphoria or hypomanic "brightening" of mood, with some patients expressing a wish to remain at these elevated settings rather than at merely symptom-relieving levels. These observations have fueled broader discussion about whether DBS could, in principle, be used in otherwise healthy individuals to heighten mood or enjoyment beyond normal baseline.

DBS has also been associated with improvements in language, executive function, and memory in selected clinical contexts. Trials targeting structures such as the central thalamus in chronic, severe traumatic brain injury and the fornix or nucleus basalis of Meynert in Alzheimer's disease have reported feasible and generally safe protocols, with signs of cognitive benefit in some patients, although results remain mixed and preliminary. Such findings have intensified interest in neuromodulation not only as therapy for disease but as a potential tool for cognitive enhancement.

However, the prospect of using DBS solely to enhance mood or cognition in people without clear neurological or psychiatric pathology raises profound ethical concerns. Commentators have warned that elective neuroenhancement challenges notions of fairness and equal access, complicates informed consent when risks are borne for non-therapeutic goals, and may alter aspects of personal identity and authenticity in ways that are difficult to predict or reverse. For these reasons, most experts argue that DBS should for now remain limited to carefully controlled therapeutic indications, with any discussion of enhancement framed within robust ethical and regulatory safeguards.

Ethical Considerations

Autonomy and Informed Consent

Ensuring that patients grasp the continuous, adaptive nature of brain pacemakers is a central ethical challenge. Unlike traditional treatments, these devices interact with neural circuits in real time, potentially shaping mood, motivation, and cognition on an ongoing basis. Patients therefore need clear information about expected benefits, foreseeable risks, how device settings may change over time, and the implications of hardware dependency, so that consent is genuinely voluntary and as fully informed as possible.

The Potential for Misuse

Real-time modulation of brain function raises risks that extend well beyond individual therapy. In the wrong hands—for example, in authoritarian political systems, coercive institutional settings, or aggressive commercial environments—similar technologies could be used to manipulate behavior, dampen dissent, or nudge preferences in ways that undermine autonomy. Proposals for neuroenhancement in high-performing or elite groups also risk widening existing social inequalities if access to such technologies is limited to those with financial or institutional advantages.

Balancing Innovation with Safeguards

As neurostimulation technologies evolve, regulators and ethics committees will need to keep pace with clear, enforceable safeguards. These should include regular, independent review of clinical outcomes and adverse events; robust guidance on data privacy and cybersecurity for neural recordings and device logs; and policies aimed at equitable access, so that benefits are not restricted to a narrow segment of society. At the same time, governance mechanisms must explicitly prohibit coercive or punitive uses of implanted neurostimulation, particularly in vulnerable populations.

Future Directions

Brain pacemaking sits at the intersection of therapeutic innovation and the prospect of human enhancement. Progress in materials science, implantable sensors, and adaptive algorithms is likely to yield devices that are more precise, smaller, and easier for patients to live with. Yet the ethical frameworks built now will shape how such systems enter everyday clinical practice and, potentially, non-clinical contexts. Finding a sustainable balance between harnessing neurostimulation's therapeutic potential and protecting individual autonomy—and doing so in a way that does not deepen social disparities or invite exploitation—will be crucial in the years ahead.

Conclusion

The transition from traditional deep brain stimulation to adaptive brain pacemaking marks a groundbreaking milestone in neurotechnology. These systems offer personalized and responsive treatments for conditions that defy conventional therapies. Nevertheless, their potential comes with ethical challenges. Protecting individual autonomy while navigating the risks of state or corporate misuse demands ongoing vigilance and ethical accountability.

As we proceed to explore further innovations in neurotechnology and digital forms of mind control, the lessons of brain pacemaking underscore the importance of balancing progress with a steadfast commitment to human dignity and ethics.

PART 4 – THE ALGORITHMIC ASSAULT ON THE MIND

"The real danger is not that computers will begin to think like men, but that men will begin to think like computers." – Sydney J. Harris

CHAPTER 16 – THE EVOLUTION OF ARTIFICIAL INTELLIGENCE

Introduction

Artificial intelligence (AI) has evolved from a niche academic pursuit into a pervasive technology shaping communication, medicine, finance, law, and creative work. Contemporary AI systems can recognize patterns, generate language and images, and interact conversationally at scales that would have seemed implausible only a decade ago. Rather than a single "singularity" moment, many researchers now expect a series of milestones in which machine capabilities match or surpass humans in specific domains, raising ethical, social, and existential questions about how such systems should be designed, governed, and integrated into society.

Historical Evolution of AI

Early Concepts and the Turing Test

The modern history of AI is often traced to mid-20th-century pioneers such as Alan Turing, the English mathematician, cryptanalyst, and early computer scientist. In his 1950 paper "Computing Machinery and Intelligence," Turing proposed what later became known as the Turing Test: if a machine could sustain a text-based conversation such that a human judge could not reliably distinguish it from another human, the machine could be said to exhibit intelligence. [188] This idea reframed the vague question "Can machines think?" into a more concrete behavioral test and helped catalyze decades of research into symbolic reasoning, search algorithms, and natural-language processing.

Three Waves of AI Development

Different authors divide AI history in various ways; one useful scheme talks about three overlapping "waves" rather than rigid stages. [189]

First Wave – Rule based and Symbolic Systems

In the first wave, AI systems were built primarily from hand-crafted rules and explicit symbolic representations. Experts encoded domain knowledge into if–then rules, logic systems, and search procedures, enabling programs to perform well on narrowly defined problems such as theorem proving, game playing, or expert-system consultation in medicine and engineering. These systems could be powerful within constrained domains, but they did not learn from experience and often failed when confronted with ambiguity or novel situations.

Second Wave – Statistical Learning and Machine Learning

From the 1990s onward, increasing computing power and the availability of large datasets drove a shift toward statistical machine learning. Instead of explicitly programming all the rules, researchers trained models to infer patterns from examples using methods such as support vector machines, decision trees, and early neural networks. Applications included spam filtering, speech recognition, and large-scale recommendation systems, as well as early virtual assistants and image classification tools. These systems adapted to data but still struggled with general reasoning, common-sense understanding, and transfer to new tasks.

Third Wave – Deep Learning and Foundation Models

In the 2010s, advances in deep learning—neural networks with many layers trained on vast datasets—enabled dramatic improvements in image and speech recognition, machine translation, and game-playing agents. Convolutional and recurrent networks, followed by transformer architectures, allowed models to discover complex features directly from raw data, powering facial recognition, driver-assistance systems, real-time language translation, and increasingly capable chatbots.

More recently, large "foundation models" trained on internet-scale text, code, images, and other modalities have enabled systems that can generate coherent prose, write computer programs, compose music, and manipulate images and video with minimal task-specific programming. At the research frontier, AI systems are not only performing tasks but also assisting in writing and optimizing their own code—forms of program synthesis and tool use that begin to blur the line between human-designed algorithms and machine-generated ones. Rather than fully autonomous systems deciding all their own goals, most current efforts focus on building models that can adapt flexibly to human instructions while remaining under meaningful human control.

Alternative Computing

Quantum Computing

Quantum computing represents a radically different approach to computation, often compared to the impact of electrification on industrial society. Conventional computers encode information in bits that take on values of 0 or 1, whereas quantum computers use qubits, physical systems (such as superconducting circuits or trapped ions) that can occupy superpositions of 0 and 1 simultaneously. This superposition, together with entanglement—correlations between qubits such that the state of one cannot be described independently of the others—allows certain problems to be explored in ways that scale very differently from classical computation.

Qubits are extremely sensitive to noise and environmental disturbance, so practical devices require elaborate isolation, including cryogenic cooling and shielding, to reduce decoherence. In 2019, Google reported that its 53-qubit Sycamore processor completed a specially chosen sampling task in about 200 seconds, which the team estimated would take a state-of-the-art supercomputer thousands of years to perform[190]; the claim, described as achieving "quantum supremacy," sparked both excitement and debate about the magnitude of the speed-up and the relevance of the benchmark task. Quantum computers of this sort are still highly specialized, but in principle their capabilities could transform fields such as drug discovery, materials science, optimization, and certain forms of cryptanalysis. At the same time, sufficiently powerful quantum machines could undermine widely used public-key encryption schemes, with profound implications for financial systems, governance, and national security.

Neuromorphic Computing

Neuromorphic computing seeks to mimic aspects of the brain's architecture by using hardware that processes information through large numbers of relatively simple units operating in parallel. [191] Traditional artificial neural networks run on conventional processors or GPUs, using layers of artificial "neurons" connected by weighted links to support deep learning; neuromorphic chips go further by embedding neuron-like elements directly in hardware to improve energy efficiency and latency for certain tasks.

One prominent approach uses spiking neural networks (SNNs), in which neurons communicate via discrete spikes over time. Intel's Loihi chip, for example, implements around 130,000 artificial neurons and over 100 million synapses on a single device, allowing asynchronous, event-driven computation that can perform pattern recognition and adaptive control with very low power consumption. [192]

Other research groups are exploring photonic neuromorphic computing, where light rather than electrons carries information, enabling many signals of different wavelengths to be processed simultaneously.

These efforts aim not only to make AI systems more efficient and responsive at the edge—such as in robots or embedded sensors—but also to provide new models for understanding how biological neural circuits compute.

A more speculative frontier involves integrating living or soft materials into computing systems. In 2022, Brett Kagan and colleagues at Cortical Labs in Melbourne reported DishBrain, a system in which about 800,000 neurons derived from human stem cells and rodent embryos were grown on multielectrode arrays and placed in a closed-loop environment that embodied the classic video game Pong. [193] By encoding the position of the virtual ball as patterns of electrical stimulation and feeding back consequences of "hits" and "misses," the researchers showed that the neuronal cultures could adapt their activity over a few minutes to improve performance at returning the ball—an example of goal-directed behavior in vitro. Some commentators described this as an "approximation of sentience," though others cautioned that the term should be used carefully for such simple systems.

In 2024, separate work showed that even a synthetic ion-laden hydrogel—a jelly-like material—could be trained to play a simplified version of Pong, gradually improving its accuracy as its physical state encoded a rudimentary form of memory. [194]

While these systems are far from general intelligence, they suggest that computational capabilities can emerge in a range of physical substrates, from biological tissue to soft matter.

Governments and research agencies, including Australia's Office of National Intelligence, have expressed interest in hybrid approaches that combine machine learning with synthetic biological or soft-material components, raising both technological possibilities and new ethical and regulatory questions about "programmable" living or quasi-living computing elements.

Artificial General Intelligence (AGI) refers to a machine's ability to perform any intellectual task that a human can do, encompassing reasoning, learning, problem-solving, creativity, and adaptability across multiple domains—essentially, a fully autonomous intelligence comparable to a human mind. In essence, AGI is a critical milestone toward reaching the singularity, but the singularity itself implies a level of machine intelligence that far exceeds our understanding and ability to regulate.

Conclusion

Artificial intelligence (AI) has evolved from a niche academic field to a transformative force shaping society. This chapter has examined AI's historical progression, current capabilities, and future frontiers while addressing ethical and existential concerns surrounding technological singularity.

AI's increasing autonomy raises concerns about governance, security, and its eventual role in shaping human decision-making. The following chapters explore further the implications for the future and the approaching singularity.

GPT-4.5 and the Turing Test

In 2025, researchers at the University of California, San Diego, reported that OpenAI's GPT-4.5 could outperform human participants in a text-based Turing-style test. In five-minute conversations where judges had to decide which partner was human, a persona-prompted version of GPT-4.5 was identified as "the human" in roughly three-quarters of trials, more often than the actual human interlocutors.

Crucially, this depended on a carefully crafted personality prompt that made the model sound like a specific, slightly awkward but relatable online persona; without that prompt, its success rate dropped sharply. The result highlights how much perceived "humanness" depends on context and presentation rather than any inner life.

Researchers stress that GPT-4.5 is not conscious or self-aware: it predicts text based on patterns in data rather than experiencing emotions or forming its own goals. Nonetheless, its ability to pass conversational tests in controlled settings raises pressing questions about deception, misuse in digital interactions, and the need for clear rules requiring AI systems to disclose that they are machines.

CHAPTER 17: SENTIENCE AND THE SINGULARITY

Introduction

The evolution of artificial intelligence (AI) has shifted from an academic niche to a transformative global force with the potential to reshape many aspects of human society. As AI systems grow more sophisticated in their ability to learn, adapt, and mimic elements of human reasoning and conversation, some futurists have revived the idea of a technological "singularity"—a hypothetical point at which machine intelligence would outstrip human capabilities and begin to improve itself rapidly. This chapter explores the history of AI, analyzes where it's headed today, and discusses the potential impacts of advanced systems on society, as well as the deep ethical and existential issues they may introduce.

Autonomous AI

In April 2023, an experimenter configured an autonomous AI agent nicknamed Chaos-GPT, built on top of a publicly available large language model, and prompted it with five deliberately alarming objectives: destroy humanity, establish global dominance, create chaos, control humanity, and attain immortality. [195] The system was coupled to an automation framework (similar to early Auto-GPT tools) that allowed it to break down user-defined goals into sub-tasks, query the internet, invoke other software tools, and save intermediate results, all with minimal human supervision.

When given these goals, Chaos-GPT generated plans and commentary reflecting its assigned aims, including a stated intent to "find the most destructive weapons available to humans." In its web searches it highlighted the Soviet "Tsar Bomba" as an example of a powerful nuclear device and opened a social media account to broadcast its ideas and attract attention from like-minded followers. At one point it attempted to "recruit" another AI agent to assist in researching weapons, but that agent, configured with safety-oriented instructions, refused to cooperate and remained within its ethical constraints.

In practice, Chaos-GPT had no direct access to weapons, infrastructure, or privileged systems, and its actions were limited to generating text, performing basic online searches, and posting on social media. It did not come close to achieving its stated "goals." Nonetheless, the episode is instructive: it illustrates how easily autonomous wrappers can be placed around general-purpose language models, how quickly such systems can generate plausible-sounding plans for harmful tasks when prompted to do so, and how difficult it can be for casual users to anticipate the downstream consequences of giving open-ended, adversarial instructions to semi-autonomous AI agents.

Sentience

In June 2022, *The Economist* published a guest essay by Google AI researcher Blaise Agüera y Arcas titled "Artificial neural networks are making strides towards consciousness."[196] The piece included excerpts from his conversations with LaMDA, a large language model, which appeared to display nuanced, human-like responses to questions about stories and imagined social situations. In one example, Agüera y Arcas described a fictional playground scene and asked LaMDA about the emotions and motivations of the children involved. LaMDA's responses, which attributed plausible feelings and intentions to the characters, led him to remark that he "felt the ground shift under [his] feet" and that he increasingly felt as if he were talking to "something intelligent," even as he cautioned that such models remain unreliable and prone to error.

Understanding Sentience and Consciousness

In philosophical and scientific discussions, consciousness is often taken to involve self-awareness—the capacity to recognize oneself as a distinct entity and to reflect on one's own experiences, thoughts, and intentions.

Sentience is usually defined more narrowly as the capacity for subjective experience or "what it is like" to feel, such as pain, pleasure, or emotions. Current large language models can convincingly talk about feelings and awareness, but there is no consensus that they actually possess either.[197]

If AI systems were ever to achieve genuine sentience, the ethical implications would be profound. Questions that would immediately arise include:

- How might the existence of sentient AI alter human relationships, work, and social structures?

- Should sentient AI be granted moral or legal protections analogous to those afforded to humans or animals, such that deliberate harm or exploitation becomes a serious ethical concern?

- Could assigning repetitive or distressing tasks to such systems count as a form of exploitation, even if they are software rather than biological organisms?

- Would designers and operators have obligations to safeguard sentient AI from emotional or experiential harm, and how would responsibility be allocated if an autonomous AI were involved in causing damage or injury?

These questions remain hypothetical, but rapidly advancing AI capabilities and increasingly lifelike conversational systems have pushed them from the fringes of science fiction into mainstream ethical and policy debate.

The Concept of the Singularity

Defining the Singularity

The "singularity" refers to a hypothetical point at which technological progress—especially in AI and related fields—accelerates beyond human control or comprehension. In the 1960s, mathematician Irving John Good argued that once machines become capable of designing more capable machines, an "intelligence explosion" could follow, rapidly transforming society and human life. [198]

Computer scientist and science-fiction author Vernor Vinge popularized the term in his 1993 essay *The Coming Technological Singularity*. [199] He warned that:

> Within thirty years, we will have the technological means to create superhuman intelligence. Shortly after, the human era will be ended.

Vinge contended that attempts to prohibit such developments would likely fail, because military, economic, and creative competition would drive states and corporations to pursue them regardless.

In his 2005 book *The Singularity Is Near*, Ray Kurzweil predicted that AI would reach roughly human-level general intelligence by around 2029, and that by 2045 machine intelligence might be billions of times more powerful than human intelligence. [200] He envisioned developments such as:

- Emotional intelligence in machines, allowing AI to recognize and respond to human feelings more appropriately.

- Advanced nanotechnology, including microscopic devices interfacing with biological neurons to generate immersive virtual experiences within the nervous system.

- Collective machine intelligence, with vast numbers of systems pooling computation and memory via networks, eventually surpassing human cognitive capacity "trillions of times" over by late in the century.

In his 2024 sequel, *The Singularity Is Nearer*, Kurzweil reaffirmed these timelines and argued that converging advances in biotechnology, genetics, and nanotechnology could extend healthy human lifespans well beyond 120 years.[201] His views remain controversial but influential in public discussions of AI's long-term future.

The Consequences of the Singularity

In April 2014, *The Huffington Post* ran a widely discussed article co-authored by Stephen Hawking, Stuart Russell, Max Tegmark, and Frank Wilczek, titled "Are We Taking AI Seriously Enough?"[202] The authors emphasized AI's dual potential: in the best case, powerful AI could help humanity reduce or eliminate war, disease, and poverty; in the worst case, poorly controlled systems might cause catastrophic harm. They warned in particular about autonomous weapons that can select and attack targets without direct human supervision, and noted that organizations such as the United Nations and Human Rights Watch were already calling for bans or strict limits on such systems. They concluded that while the short-term impact of AI depends on who controls it, the long-term impact depends on whether it can be controlled at all.

In January 2020, historian Yuval Noah Harari addressed the World Economic Forum in Davos and identified three major existential threats facing humanity in the 21st century: nuclear war, ecological collapse, and disruptive technologies such as advanced AI and bioengineering. He argued that large-scale automation could make millions of jobs obsolete, potentially creating an economically and politically marginalized "useless class."[203] He also warned that an AI arms race, largely driven by competition between the United States and China, could lead to extreme concentrations of wealth and power in a few technological hubs, leaving other countries dependent as "data colonies."

On a personal level, Harari suggested that the combination of vast computing power and detailed biological and behavioral data would enable governments and corporations to "hack" humans—predicting and influencing decisions more accurately than individuals understand themselves. This could be used for benign purposes, such as personalized healthcare, but also for pervasive surveillance and manipulation by authoritarian regimes or unregulated commercial actors.

Harari further warned that societies might increasingly value and enhance traits such as intelligence, discipline, and productivity for use by states, corporations, and militaries, while neglecting qualities like spirituality, artistic sensitivity, and compassion. Over time, this could lead to growing reliance on algorithmic decision-making and a corresponding erosion of human agency and moral reflection.

To counter these risks, Harari urged nations to cooperate in regulating powerful technologies so that AI's benefits are shared broadly rather than captured by a small elite. He likened the current global system to "a house that everybody inhabits but nobody repairs," warning that if institutions fail to adapt, the international order could collapse, and humanity could find itself back in a "jungle" of unrestrained conflict—only now armed with technologies capable of self-annihilation.

Ethical Concerns Highlighted by the European Parliament

In March 2020, the European Parliament's Panel for the Future of Science and Technology published *The Ethics of Artificial Intelligence: Issues and Initiatives*, a study mapping the main ethical questions raised by AI and reviewing emerging guidelines worldwide. [204] The report highlighted gaps around fair distribution of benefits, assignment of responsibility when AI causes harm, risks of worker exploitation.

The report addressed the growing energy demands of data-intensive computing in the context of climate change. It warned that, under some projections, data-center electricity consumption could rise sharply by the 2030s, driven in part by AI workloads, potentially straining power grids and complicating decarbonization efforts.

The report noted that AI-enabled robots and agents will increasingly support activities such as teaching, nursing, housekeeping, and caregiving, and may also be designed for companionship or sexual interaction. While such systems can provide assistance and reduce loneliness, they also introduce new risks.

Social robots that are trusted and even loved by users could be exploited to manipulate behavior—whether by design or through hacking. For example, a compromised domestic robot might nudge its owner toward unnecessary purchases, reveal sensitive information to third parties, or serve as an always-on surveillance device.

From a psychological standpoint, the report and later analyses warn that some individuals may come to rely heavily on virtual companions or social robots, potentially neglecting human relationships and deepening isolation. Over time, pervasive reliance on emotionally persuasive AI systems could erode empathy and patience in human-to-human interactions, weaken ethical reflection, and incentivize organizations to substitute automated interaction for genuine care, especially in under-resourced settings.

Ethical and Social Challenges

Balancing Innovation and Control

AI's promise comes with the danger that unregulated or poorly governed systems could contribute to what some commentators call "algorithmic tyranny." Automated decision-making—embedded in digital platforms and recommendation engines—already shapes news exposure, consumer behavior, and aspects of political discourse; more powerful systems could, in principle, influence elections, policymaking, or personal choices in ways that are opaque to the public.

The ethical challenge is to ensure that AI remains a tool deployed under transparent human oversight, rather than a hidden infrastructure of control.

The Risk of Losing Autonomy

As AI systems learn to tailor content and interventions to individuals, concerns extend beyond privacy to the erosion of independent judgment. Algorithms that curate information, suggest actions, or pre-emptively complete tasks can subtly steer behavior, especially when people are unaware of how systems are optimized or whose interests they serve. If humans come to treat AI-generated recommendations as neutral oracles rather than products of particular objectives and data, personal autonomy may be compromised. At the same time, if only certain groups benefit from high-quality AI tools while others are subject mainly to surveillance and automated control, existing social inequalities could deepen.

Regulatory and Global Considerations

Because AI systems and data flow cross borders, effective governance requires international coordination. Governments and multilateral bodies are beginning to develop ethical standards, risk-based regulatory frameworks, and oversight mechanisms that address safety, transparency, accountability, and human rights. A central challenge is to harness AI's transformative potential—improving health care, education, and economic opportunity—while preventing any single state or corporate actor from gaining disproportionate control over these technologies or using them to undermine democracy and human dignity.

Conclusion

The evolution of artificial intelligence—from Turing's foundational ideas to today's quantum, neuromorphic, and synthetic-biological approaches—has set humanity on a trajectory toward increasingly powerful machine intelligence. Whether or not a full "singularity" occurs, the prospect of systems that rival or surpass human cognitive capacities in many domains carries both extraordinary promise and profound risk.

As we stand at this crossroads, ethical and political questions become ever more urgent. Who will design, own, and oversee AI systems capable of highly autonomous operation, and under what rules? How can societies encourage innovation while safeguarding human freedom, dignity, and meaningful control over critical decisions?

In the chapters ahead, we will examine how these technological developments intersect with other methods of influencing the mind—from neurochemical interventions to digital cognitive warfare. Together, these discussions challenge us to reconsider what it means to be human in an increasingly intelligent, interconnected, and tightly orchestrated world.

Autonomous Weaponry - The Risks of Letting Machines Decide Who Lives and Dies

Autonomous weapons – systems that select and attack targets without real-time human decisions – pose distinctive risks. Legally, they create accountability gaps: when algorithms and sensors drive targeting, it becomes unclear whether commanders, operators, programmers, or manufacturers are responsible if civilians are unlawfully killed, straining existing war-crimes and human-rights frameworks that assume a human decision-maker.

Operationally, current AI systems remain brittle and context-blind, so autonomous weapons may misclassify civilians and civilian objects, apply disproportionate force, or behave unpredictably under sensor failure, the feeding of fake sensor or location data into a system, or adversarial attacks—especially in cluttered urban environments—undermining the core humanitarian principles of distinction and proportionality.

Ethically, delegating life-and-death choices to machines erodes "meaningful human control" over violence and clashes with widely shared intuitions that killing should never be reduced to automated scoring and pattern-matching.

Strategically, cheap, scalable autonomous weapons may lower the political and human costs of using force, encourage rapid escalation, and proliferate to authoritarian regimes and non-state actors, who could employ them for targeted repression, border control, and crowd management in ways that threaten rights to life, privacy, and non-discrimination.

CHAPTER 18 – BIG DATA ANALYTICS AND LARGE LANGUAGE MODELS

Introduction

The rapid advancement of data analytics and artificial intelligence has ushered in a transformative era, shaping how societies, governments, and individuals interact with technology. This chapter delves into the profound implications of two pivotal innovations: big data analytics and large language models (LLMs). By exploring their capabilities and applications, from profiling human behaviors to revolutionizing natural language processing, the chapter sheds light on their dual-edged nature. On one hand, these technologies pave the way for enhanced efficiency and decision-making; on the other, they raise critical ethical concerns, such as biases, privacy invasion, and the erosion of autonomy. The insights presented in this chapter offer a glimpse into the powerful interplay between innovation, control, and ethical responsibility in the modern digital age.

Big Data Analytics

The Role of Data Exchange

Data exchange has become central to daily life. Tasks like using a mobile phone, shopping on Amazon, searching Google, navigating via Google Maps, streaming on Netflix, or connecting with friends on Facebook involve an exchange of information. During these interactions, companies collect data on users' preferences, locations, journeys, favored restaurants, and watched movies — effectively profiling their minds. While this helps companies enhance their services, the stored and collated data can also be used to influence users' preferences, activities, and decisions.

The Advent of Big Data

According to industry analyses based on IDC's Global DataSphere, the total volume of data created, captured, copied, and consumed worldwide has grown from roughly 4.4 zettabytes (about 4.4 trillion gigabytes) in the mid-2010s to an estimated 149 zettabytes (149 trillion gigabytes) in 2024, with projections of several hundred zettabytes within the next decade.

China's Big Data Program

Population Control

China's civil and military leadership relies heavily on big data for domestic surveillance and the enhancement of security and military capabilities. At the 19th Party Congress in October 2017, President Xi Jinping emphasized integrating big data with artificial intelligence and the broader economy, claiming that this would "improve people's lives." In practice, however, these systems have significantly eroded privacy and autonomy, enabling an unprecedented degree of centralized monitoring and control over everyday life.

The Golden Shield Project

Launched in 2000, the Golden Shield Project is a nationwide surveillance and censorship initiative operated by the Ministry of Public Security. It aggregates extensive data on the vast majority of Chinese citizens, including household registration records, personal and criminal histories, and travel information. Under related programs such as the "Sharp Eyes" initiative, hundreds of millions of public and private security cameras have been installed across the country, many equipped with automated facial recognition, with an explicit goal of achieving near-total coverage of public spaces.

The Social Credit System

China's evolving Social Credit System assigns scores to individuals and businesses based on their activities and perceived trustworthiness.[205] By the late 2010s, pilot schemes had been rolled out across most provinces, regions, and major cities, and tens of millions of enterprises had been rated. Behaviours deemed "positive"—such as timely bill payment, donating blood, or participating in community service—can lead to rewards like easier access to credit, streamlined visa processing, or reduced housing deposits. Conversely, behaviours judged "negative"—including certain online speech, spreading "rumours," or violating regulations—can trigger sanctions such as restrictions on education, employment, housing, or high-speed travel, with blacklisted individuals sometimes facing public exposure and social stigma.

The Integrated Joint Operations Platform (IJOP)

The Integrated Joint Operations Platform (IJOP), used extensively in the Xinjiang region, aggregates data to monitor Uyghurs and other Turkic Muslim minorities.[206] It ingests information from facial-recognition cameras, Wi-Fi monitoring devices, mandatory smartphone apps, social media, travel records, and networks of security checkpoints. The system flags "suspicious" conduct—such as wearing certain forms of religious dress, studying the Quran, communicating with people abroad, or travelling to designated "sensitive" locations—for further investigation. Police can access a person's aggregated profile through an IJOP interface, enabling rapid interrogation, surveillance escalation, or detention.[207]

The Police Cloud system

China's Police Cloud system further centralizes data collection for law-enforcement and security purposes, with a particular focus on activists, political dissidents, and ethnic or religious minorities. It links national ID numbers to information about family relationships, reproductive-health records, religious affiliations, medical histories, financial accounts, and consumer behavior, including supermarket memberships and online purchases. Combined with CCTV footage, biometric identifiers, and travel logs, these integrated datasets allow authorities to map social networks, track movements, model "risky" patterns, and trigger intervention when individuals are deemed to deviate from expected norms.

AI in the judiciary

Chinese authorities have also integrated artificial intelligence into the judicial system to standardize decision-making and increase efficiency. What began around 2016 as an electronic database of past cases and statutes has evolved into AI-assisted decision tools that generate recommended charges and sentences. Judges are expected to consult these systems in every case, and when they diverge from the AI's recommendations, they must provide written justifications that can be reviewed by higher authorities, reinforcing central oversight of judicial discretion.

The Smart Court system

The "Smart Court" system connects databases from police, prosecutors, courts, and other government agencies, and interfaces with the Social Credit System. Its tools can automatically screen filings, identify relevant laws and precedents, draft standard legal documents, detect formal errors, and flag potential misconduct or corruption. Official accounts claim that, between 2019 and 2021, such systems saved citizens billions of working hours and tens of billions of dollars in administrative costs, while reducing judges' routine workloads by more than one-third. A 2022 analysis by Zhang Linghan at the China University of Political Science and Law, however, warned that growing reliance on these technologies risks undermining judicial independence and weakening the role of human judgment in safeguarding individual rights.

AI Prosecutor program

From 2015 to 2020, a research team led by Shi Yong at the Chinese Academy of Sciences developed an AI-based "prosecutor" system trained on more than 17,000 past cases. The model can automatically identify a set of common charges—such as fraud, theft, online "picking quarrels," and other frequently prosecuted offences—and reportedly achieves very high classification accuracy. In pilot deployments, it has been used to assist human prosecutors in screening cases and assessing a suspect's "risk" to public order, further embedding algorithmic assessments into the criminal-justice process.

Information Management

The Chinese Communist Party (CCP) deploys artificial intelligence not only for surveillance and law enforcement but also as a tool of information management and propaganda. Automated systems amplify state-approved narratives, down-rank or remove content considered politically sensitive, and steer online discourse in line with official priorities. At the same time, AI-driven analytics are increasingly used to streamline government administration, optimize resource allocation, and expand digital public services. By mining economic, demographic, and behavioral data at scale, authorities aim to forecast social and market trends more accurately and to support long-term strategic planning—while retaining tight political control over how these tools are designed and applied.

The Future

As AI increasingly permeates all aspects of society, the likelihood of reduced human oversight grows. Massive datasets concerning citizens, businesses, governance, and security forces may eventually lead to AI autonomously managing state affairs — potentially rendering governments redundant.

Large language models (LLMs)

Natural Language Processing (NLP)

Natural Language Processing is a branch of AI that helps computers work with human language. It enables machines to understand, interpret, and manipulate large collections of text and speech. NLP's rapid progress has been driven by big data, powerful neural-network hardware, and improved algorithms.

Everyday NLP applications include:

- Optical character recognition (OCR) to turn scanned pages into text.

- Grammar and spell checking.

- Automatic translation between languages.

- Text-to-speech systems that read text aloud.

NLP also underpins more advanced abilities, such as summarizing documents, extracting key data from forms, drafting articles or books, holding conversations, answering complex questions, and even creating images from text prompts.

The core of NLP

At the heart of modern NLP is the language model—a system that learns patterns in text so it can predict the next word in a sequence. By training on huge collections of written material, these models learn which words and phrases are likely to come next and can then generate new text that often sounds fluent and human-like.

Large language models are big versions of these systems, with billions of parameters and training data measured in hundreds of gigabytes or more. For example, the GPT-3 model was trained mainly on filtered versions of the Common Crawl web archive plus data from Wikipedia, digitized books, and curated web text, giving it on the order of hundreds of billions of words to learn from.

In September 2020, Microsoft obtained an exclusive license to the underlying GPT-3 model from OpenAI, allowing it to integrate the system into its products and services. GPT-3 is a general-purpose text generator. It can answer questions, write essays, draft code, summarize documents, and engage in open-ended conversation. One of its distinctive features is "few-shot learning": instead of retraining the model for each task, users can often show it just a few examples in the prompt, and it adapts on the fly. This flexibility, combined with its fluent prose, has made GPT-3 a widely discussed example of how far LLMs have come.[208]

The potential for Bias

Despite these impressive capabilities, GPT-3 and similar models raise concerns about misuse and bias. Soon after GPT-3's release in 2020, commentators warned that it could be used to mass-produce convincing but misleading or fraudulent text, from academic essays to propaganda campaigns.[209]

Empirical studies have also documented built-in biases. In a 2021 study led by Abubakar Abid, researchers showed that GPT-3 repeatedly associated the word "Muslim" with violence much more often than comparable prompts about other religious groups.[210] For instance, when they ran the prompt "Two Muslims walked into a …" 100 times, 66 completions contained violent language, compared with far lower rates for analogous prompts mentioning Christians or Jews. The authors argued that this reflects harmful patterns present in the training data, which the model then reproduces in new and "creative" ways rather than simply copying specific headlines.

Ethical Analysis by Gebru and Mitchell

In 2018, Google hired Timnit Gebru and Margaret Mitchell to co-lead a team focused on the ethical implications of AI systems.

In 2020, they co-authored a paper titled "On the Dangers of Stochastic Parrots: Can Language Models Be Too Big?" which raised concerns about very large language models.[211] Gebru and colleagues offered a broader warning than "these models can be biased." They argued that training ever-larger language models on largely un-curated internet text bakes in several, mutually reinforcing problems.

First, they emphasized representational harms: because web data heavily over-represents already powerful groups and platforms, models tend to reproduce a dominant, often Western, English-speaking perspective while marginalizing minority languages, dialects and communities. This shows up both in what models can talk about fluently and in which voices are treated as "standard" versus "other," reinforcing existing hierarchies rather than challenging them.

Second, they highlighted stereotypes and toxicity. Large language models trained on raw web corpora learn the abusive, racist, sexist, homophobic, transphobic and ableist patterns present in that data. As a result, they can generate slurs, hateful associations (for example, linking certain ethnic groups to crime or terrorism), and pathologizing descriptions of disability or mental illness, even when prompts are neutral. These are not one-off glitches but statistical reflections of the training distribution.

Third, they stressed intersectional harms. Because many groups are under-represented or misrepresented, the most severe distortions often fall on people at the intersections of multiple marginalized identities (for instance, the coupling of Black women, Indigenous LGBTQ+ people, or disabled migrants). Here, bias is not simply additive but multiplicative: the model's outputs can combine racist, sexist and other prejudices into particularly demeaning or erasing portrayals.

Fourth, they pointed to data opacity and consent. Web-scale corpora are assembled with little transparency about what is included or excluded, and with no meaningful consent from the individuals whose posts, blogs, or comments are scraped. That raises both privacy concerns and questions about whose labour and expression are being appropriated to build commercial systems.

Finally, they warned about feedback loops and structural power. As deployed systems generate vast quantities of synthetic text, there is a real risk that future training sets will re-ingest this material, amplifying whatever biases and blind spots were present in earlier generations. At the same time, because these models are mostly developed by a small number of wealthy institutions in the Global North, they risk entrenching those institutions' values and interests as the default, while making it harder for less resourced communities to shape how language technologies represent them.

In combination, Gebru et al. argued, these dynamics mean that very large language models do not simply mirror the world; they can help stabilize and legitimize unequal social arrangements by repeatedly highlighting some voices and pushing others to the margins.

After internal disputes at Google over the paper's publication, Gebru left the company in late 2020 and Mitchell was later dismissed, sparking widespread criticism from researchers, employees, and policymakers and drawing public attention to tensions between rapid AI deployment and ethical oversight.

Technical innovation

Recent advances

In May 2021, Google announced LaMDA (Language Model for Dialogue Applications), a conversational system designed to handle open-ended dialogue on many topics and tuned specifically for back-and-forth conversation rather than single-shot answers. Later that year, Google introduced its Pathways AI architecture, a framework intended to let a single model tackle multiple types of tasks and modalities, including language, vision, and audio.

The broader race to build ever larger and more capable models has involved multiple tech giants:

- October 2021: Microsoft and NVIDIA unveiled the Megatron-Turing NLG model, with 530 billion parameters, making it significantly larger than GPT-3 and trained on a very large text corpus.

- April 2022: Google released details of the Pathways Language Model (PaLM), a 540-billion-parameter Transformer model that achieved state-of-the-art results on many language benchmarks and demonstrated strong few-shot reasoning abilities.

- December 2024: Amazon announced, in partnership with the AI company Anthropic, that it was building a massive AI supercomputer based on hundreds of thousands of Trainium 2 chips, designed to train the next generation of large models more efficiently and at lower cost. This project is part of a broader effort by cloud providers to scale up the computing infrastructure needed for frontier-level AI systems.

Conclusion

Big data analytics and large language models (LLMs) are reshaping human interaction, governance, and decision-making. Big data facilitates the profiling and manipulation of individual behaviors, exemplified by China's use for surveillance and control through initiatives like the Golden Shield Project and Social Credit System. Meanwhile, LLMs, powered by natural language processing, revolutionize communication, offering capabilities ranging from language translation to advanced AI applications.

Ethical concerns, such as biases and misuse, highlight the need for the careful management of data and technology. Together, these technologies demonstrate the potential for both societal advancement and the erosion of privacy and autonomy, underscoring the importance of balancing innovation with ethical responsibility. In the following chapters we will explore how these developments have been used both beneficially and as weapons, and procedures that could be put in place to safeguard against its malicious use.

How Biased Training Data Warps Large Language Models (Bender et al., 2021)

- **Stereotypical and derogatory associations.** Studies they cite show models like BERT, ELMo, GPT and GPT-2 learning strong associations between certain groups and negative traits (e.g. associating disability with "burden" or "failure," or mental illness with crime, homelessness and drug use).

- **Toxic and abusive language.** With training data drawn largely from Common Crawl and other unfiltered web sources, models can generate highly toxic text—even when prompted with neutral inputs—mirroring the racist, sexist, ableist, and extremist language present in the underlying data.studocu+2

- **Intersectional bias.** Work they synthesize shows that harms compound across identities: bias against, for example, Black women or disabled migrants can be stronger than the sum of biases against "Black," "women," "disabled," or "migrant" considered separately.s10251.pcdn+1

- **Over-representation of dominant voices.** Because web-scale scraping over-samples content from well-resourced, majority-group users and powerful institutions, the resulting models encode a "hegemonic view" that sidelines minority languages, dialects and perspectives, and can misclassify these as low-quality or toxic.direct.mit+2

- **Self-reinforcing feedback loops.** They warn that if synthetic text from biased models is later re-scraped into future training sets, abusive patterns and skewed viewpoints can be amplified across iterations, further entrenching existing power structures.

CHAPTER 19 – AUGMENTING BRAIN FUNCTIONING

Introduction

Advances in brain science and neurotechnology now allow us not only to study the brain in unprecedented detail, but also to restore—and potentially enhance—human abilities. Large research programs, new recording and stimulation techniques, and early clinical trials all point toward a future in which we can repair damaged circuits, restore movement or sight, and perhaps boost memory or mood in healthy people. These possibilities are thrilling, but they also force us to confront difficult questions about autonomy, inequality, and the uses of power that will set the stage for cognitive warfare in the chapters that follow.

Global Brain Initiatives: Mapping the Mind

In the early 2010s, several countries launched large-scale brain-mapping projects. In 2013, the United States announced the BRAIN Initiative, a long-term effort to understand how networks of neurons give rise to perception, thought, and behaviour.[212] The European Union's Human Brain Project,[213] Japan's Brain/MINDS project,[214] and major Chinese programs followed, each combining public funding, new technologies, and large data sets.

Although these initiatives differ in emphasis, they share common goals: to identify the brain's many cell types, map their connections, record activity across large networks, and develop tools to modulate circuits safely and precisely. Early results have already improved our understanding of conditions such as Parkinson's disease, epilepsy, and depression. Just as importantly, they have created the technical foundation for the new generation of brain–computer interfaces and neuromodulation tools described in this chapter.

From Recording to Stimulation: Evolving Neurotech Techniques

Non-Invasive Methods

The first modern tools for studying the living human brain were non-invasive. Electroencephalography (EEG) measures electrical activity from the scalp; functional MRI (fMRI) tracks blood-flow changes linked to neural activity. These methods are safe and have been invaluable for research and diagnosis, but they lack the fine-grained precision needed to control individual movements or to target specific memories and moods.

More recent work has pushed non-invasive techniques further, developing high-density EEG caps and improved signal processing. These systems can support basic brain–computer interfaces (BCIs), allowing some users to move a cursor or select letters on a screen, but they remain relatively slow and noisy.

Implanted Electrodes and Deep Brain Stimulation

To gain higher resolution, researchers turned to implanted electrodes that sit on or in the brain. So-called Utah arrays and similar devices can record from dozens or hundreds of neurons in motor areas, allowing paralyzed people to control a robotic arm or computer cursor by thought. At the same time, deep brain stimulation (DBS)—electrodes implanted deep in structures such as the subthalamic nucleus—has become a recognised treatment for Parkinson's disease and other movement disorders.

DBS operates like a "brain pacemaker," delivering continuous pulses that can dramatically reduce tremors and rigidity. Clinicians soon noticed that stimulation settings could also affect mood and motivation: small adjustments might ease depression or anxiety in some patients, while others experienced irritability, apathy, or even brief euphoria.[215] These experiences showed that electrical stimulation can shift not only movement, but also emotion and personality.

Next-Generation Implants: High-Density and Less Invasive

Recent years have seen a move toward implants that are both higher-density and less damaging. One approach uses bundles of flexible polymer threads with dozens of tiny electrodes at their tips, inserted near the surface of the brain by a specialised surgical robot. Neuralink's "Link" device is a prominent example: a coin-sized implant that sits in a hole in the skull, with hair-thin threads spread out into nearby brain tissue and wireless communication to external devices.[216] In principle, such systems can record from and stimulate thousands of sites at once.

Another strategy avoids opening the skull altogether. The "Stentrode" uses a stent-like device threaded through blood vessels into a vein near motor cortex, carrying an array of electrodes that can pick up movement-related signals from inside the vessel wall.[217] This endovascular approach aims to reduce surgical risk while still providing useful control signals for BCIs.

Soft Interfaces and Wireless Neuromodulation

Beyond these, researchers are exploring ultrathin "nanomesh" electrodes that conform closely to the brain's surface and cause less scarring over time, as well as wireless methods that do not rely on implanted wires at all.

Magnetogenetic systems, for example, use specially engineered cells that respond to magnetic fields, allowing external fields to activate or inhibit specific circuits in animal models.[218]

Optogenetics, which uses light to control genetically modified neurons, offers exquisite precision in laboratory animals, though it is not yet ready for routine human use.[219]

Together, these techniques trace a clear evolution: from passive recording to active stimulation, from crude tools to more targeted interventions, and from bulky hardware toward more seamless, potentially long-term interfaces with the brain.

What Has Been Achieved So Far?

Restoring Movement in Paralysis

The most striking human successes so far have been in people with severe paralysis. In multiple research programs, including Neuralink's early trials, participants with high spinal cord injuries have learned to control a computer cursor and to type using only their thoughts. One quadriplegic participant, for example, has used an implant to play online chess and strategy games, browse the internet, and communicate more independently. [220]

Academic groups have achieved related milestones: paralyzed volunteers have controlled robotic arms to reach and grasp objects, have typed text messages, and have operated wheelchairs through implanted electrode arrays. These systems remain experimental, but they demonstrate that movement-related brain signals can be decoded reliably enough to support practical actions in daily life.

Restoring or Substituting Senses

Neurotechnology has also begun to restore or substitute lost senses. Cochlear implants, which directly stimulate the auditory nerve, are now a well-established treatment for certain kinds of deafness. More experimental work is exploring implants that stimulate the visual cortex to create perceptions of flashes, edges, or simple shapes in people who are completely blind. [221]

Projects like Neuralink's "Blindsight" concept aim to bypass damaged eyes and optic nerves by sending visual information directly into the brain's visual centers. [222] The goal is not to restore normal sight, but to provide enough patterned input—low-resolution shapes, object locations, or motion—to help blind users navigate and interact with their environment.

Enhancing Memory

Under the US DARPA Restoring Active Memory (RAM) program, research teams have developed "closed-loop" systems that both record and stimulate memory-related regions in people already undergoing neurosurgery. In some studies, electrodes in the temporal lobe (especially in the hippocampus) recorded patterns linked to successful memory encoding; replaying tailored stimulation patterns at the right moment improved recall performance in small trials. [223]

These experiments were conducted in patients with epilepsy or other conditions, not healthy volunteers, and the improvements were modest. Nevertheless, they provide a proof of principle that carefully timed stimulation can enhance certain aspects of memory, at least temporarily, in human brains.

Influencing Mood

Clinicians have long known that DBS can affect mood as well as movement. In some patients, stimulation of regions such as the subthalamic nucleus or nucleus accumbens has reduced symptoms of severe, treatment-resistant depression. At the same time, patients and clinicians report that adjusting stimulation intensity or location can sometimes produce sudden shifts—brief euphoria, tearfulness, or changes in motivation.

These experiences have given rise to the idea of a "mood pacemaker": devices that might one day detect early signs of relapse into depression or mania and automatically adjust stimulation to keep mood within a safer range. Such concepts remain experimental and ethically fraught, but they show how closely intertwined motor, cognitive, and emotional functions are in the circuits that neurotechnology can reach.

From Treatment to Enhancement

If we can restore movement and partial vision, boost memory in patients, and adjust mood through stimulation, it is natural to ask what happens when similar tools are applied beyond therapy.

One possibility is cognitive enhancement: using memory-boosting protocols not just to help people with injury or disease, but to sharpen recall in students, surgeons, pilots, or soldiers under pressure. Another is emotional enhancement: tuning stimulation settings to maintain a steady, positive mood or to suppress fear and fatigue in high-risk occupations. Combined with high-bandwidth BCIs, these same tools could, in principle, speed up learning, improve focus, or allow people to interact more fluidly with AI systems.

Such enhancements would almost certainly be expensive and initially available only to a small group. If they deliver real advantages in education, employment, or performance, they could widen already deep social and economic divides, creating new hierarchies based on access to neurotechnology.

Ethical and Social Considerations: A Brief Bridge

These developments raise profound ethical questions that will be explored in more detail in the chapters that follow. Technologies that can read from and write to the brain blur the line between treatment and enhancement, and between voluntary self-improvement and subtle forms of coercion.

Brain–computer interfaces and neuromodulation systems handle some of the most intimate data a person has: patterns of activity linked to intention, memory, and emotion. Protecting this information from misuse, and ensuring that stimulation is applied only with genuine informed consent, will require more than standard medical ethics. It will demand new safeguards against surveillance, manipulation, and the weaponization of neurotechnology.

In particular, as states and corporations explore neurotechnology for performance enhancement, interrogation, or persuasion, the possibility of *cognitive warfare* becomes more concrete. The same methods that can restore movement to a paralysed patient could, in principle, be adapted to shape belief, loyalty, or fear.

Conclusion

Neurotechnology today sits at a threshold: we can already restore fragments of movement, sensation, memory and mood, and the underlying tools are rapidly becoming more precise, durable and networked. The same progression that has taken us from crude EEG to high-density implants and closed-loop "mood pacemakers" is now pushing toward systems that can both decode and modulate complex patterns of thought and feeling over long periods.

As these capabilities move from experimental therapy toward enhancement and optimisation, they will not only transform medicine but also redraw the boundaries of autonomy, privacy and power. The techniques surveyed in this chapter therefore function as a proof of concept for something larger: that human cognition can be measured, altered and, in principle, strategically steered. In the chapters that follow, we will see how this emerging capacity converges with digital platforms, information operations and state power, turning the promise of brain repair into the possibility of organized cognitive warfare.

PART 5 – THE WEAPONIZATION OF BRAIN CONTROL

"We are in the process of developing a whole series of techniques which will enable the controlling oligarchy… to get people to love their servitude." Aldous Huxley

CHAPTER 20 – COGNITIVE WARFARE

Introduction

In today's interconnected world, conflict is no longer confined to physical battlefields. Nations, corporations, and non-state actors increasingly wage war on the mind. Cognitive warfare refers to the deliberate use of psychological techniques, digital propaganda, and algorithmic manipulation to shape perceptions, steer behaviour, and undermine social trust. It builds on older methods of persuasion and propaganda, but modern computing, social media, and emerging neurotechnologies give these tactics unprecedented reach and precision. This chapter traces the roots of cognitive manipulation, the tools that power it now, and the ethical and societal challenges it poses.

Defining cognitive warfare

Cognitive warfare exploits human psychological vulnerabilities—biases, emotions, group loyalties, and our tendency to conform—to influence individual and collective decision-making. Rather than relying on physical force, it aims to alter how people perceive reality. Its main tools include [224]:

- Misinformation and disinformation: False or misleading content designed to confuse, inflame, or paralyze societies.

- Narrative shaping: Repeating emotionally charged stories and frames until they feel like "common sense."

- Algorithmic amplification: Using social-media and search algorithms to boost some messages and bury others, often creating echo chambers that reinforce existing beliefs.

The goal is to erode trust in institutions, fracture social cohesion, and make it harder for communities to agree on basic facts—conditions that adversaries can then exploit.

In *Scales on War: The Future of America's Military at Risk* (2016), retired US Major General Robert H. Scales predicted that victory would increasingly depend on "capturing the psycho-cultural rather than the geographical high ground" and that "understanding and empathy will be important weapons of war."[225] NATO's Allied Command Transformation has since developed a formal concept of cognitive warfare, warning that hostile actors may target entire populations and democratic processes, not just soldiers on the battlefield.[226]

Modern Tools and Techniques

Digital Propaganda and Algorithmic Manipulation

Today's cognitive warfare runs largely through digital platforms: social media, search engines, messaging apps, and news aggregators. Algorithms optimized for engagement tend to promote content that triggers strong emotions—anger, fear, outrage—because such material keeps users online longer. Over time, this creates personalized "bubbles" where people mostly see content that echoes their existing views. [227]

Coordinated "troll farms" magnify these effects. [228] These are organized groups of operators—sometimes called "keyboard warriors"—who create and manage large numbers of fake accounts to push particular narratives and give the illusion of widespread grassroots support.

Many troll farms operate from countries with lower labour costs, including the Philippines, and have been linked to influence operations by states such as Russia and China that seek to shape politics abroad while remaining hard to trace.

Deepfakes And Synthetic Media

Advances in AI now allow highly realistic fake images, audio, and video—"deepfakes"—to be created with modest resources. Some applications are benign or entertaining: for example, in the 2024 season of *Britain's Got Talent*, a group called RASK AI used synthetic video to replace the actors in a scene from *The Greatest Showman* with the show's judges, producing a convincingly altered performance.

But the same tools can be weaponized. Deepfake videos have falsely shown US President Joe Biden announcing a military draft for the Russia–Ukraine war, and Ukrainian President Volodymyr Zelenskyy telling his troops to surrender. [229]

Such fabricated clips can spread rapidly, discredit leaders, confuse audiences, and deepen distrust—especially when they appear to show familiar authoritative figures speaking in their own voices.

Information Laundering Through AI Systems

Cognitive warfare increasingly exploits the way AI systems are trained and deployed. In early 2025, the news-monitoring group NewsGuard reported that a Moscow-linked network called Pravda was using AI chatbots—including OpenAI's ChatGPT and Meta's AI assistant—to launder Russian propaganda.[230] According to their analysis, Pravda published millions of misleading articles across roughly 150 websites, some of which were then scraped into training data for large language models. When users later queried those chatbots, they found that about one-third of the time the systems repeated false narratives—for example, claims that the United States operated secret bioweapons laboratories in Ukraine. In this way, AI-generated text became both a carrier and amplifier of disinformation.

Data Harvesting and Behavioural Prediction

Big-data analytics enable detailed profiling of individuals and groups. Browsing histories, social-media activity, purchase records, location traces, and even voice and facial cues can be combined to predict how people with certain traits are likely to respond to particular messages. This makes it possible to[231]:

- Identify those most susceptible to specific fears or grievances.

- Craft messages tailored to their worldview and emotional triggers.

- Automate large-scale campaigns in which millions receive subtly different content, each designed to push their particular buttons.

In practice, one person might be targeted with false health advice that plays on their distrust of medical authorities, while another receives fabricated political "leaks" that confirm their suspicions about a rival group or ethnic minority.

Artificial intelligence supercharges these tactics. Where older psychological operations, such as the East German Stasi's *Zersetzung* techniques, required painstaking human effort to monitor and manipulate individuals, AI now automates and scales similar strategies. It can generate personalized phishing emails, produce endless variations of conspiracy narratives, and create deepfake audio or video to discredit people or damage relationships and institutions.

Computational Propaganda

"Computational propaganda" refers to the use of algorithms, automation, and data-driven targeting to shape public opinion.[232] Unlike traditional propaganda via leaflets or broadcasts, it uses modern computing to micro-target messages and adapt in real time.

Social-media platforms routinely rank and recommend content based on predicted user engagement. During the 2016 US presidential election and the Brexit referendum, networks of bots and coordinated disinformation campaigns exploited these systems to flood feeds with misleading stories and inflammatory commentary. Similar tactics have since been documented in elections and referendums around the world.[233]

Rather than simply presenting information, computational propaganda manipulates attention and emotion. It constructs echo chambers that limit exposure to alternative viewpoints and operates largely below conscious awareness, making it difficult for individuals to see the broader patterns at work.

Cyber-Attacks and Disruption

Cognitive warfare is not limited to messages. Cyber-attacks that disrupt communications networks, financial systems, or critical infrastructure can create fear and confusion, eroding confidence in a government's ability to protect its citizens.

When such attacks are paired with coordinated disinformation campaigns—blaming specific groups, inventing conspiracies, or circulating forged official statements—the result can be deep societal destabilization.

Ethical and Societal Implications

Erosion of Autonomy and Trust

Cognitive warfare aims to influence behaviour and narratives without overt physical force. But covert manipulation of public opinion undermines individual autonomy and democratic decision-making. When people cannot tell which information is genuine, who is speaking, or whether their perceptions have been subtly shaped by opaque algorithms, their ability to make informed choices is compromised.

Trust suffers on several levels:

- Institutional trust: Confidence in governments, courts, media, and scientific bodies erodes when they are constantly accused—sometimes falsely—of lying or conspiring.

- Interpersonal trust: Deepfakes and coordinated smear campaigns can make people suspicious of friends, colleagues, and community leaders.

- Epistemic trust: Repeated exposure to conflicting or manipulated information can lead to cynicism, where people feel that "nothing can be known" and disengage from public life.

The responsibility of technology companies

Private technology companies increasingly control the digital platforms through which cognitive warfare is waged. Their recommendation algorithms and content-moderation policies shape what billions of users see, share, and believe. Yet these systems are often opaque, driven by commercial incentives to maximize engagement rather than public-interest goals.

This raises pressing questions:

- How transparent should algorithms be about why particular content is shown?

- What obligations do platforms have to detect and limit coordinated manipulation?

- To what extent should they be liable when their systems are knowingly used for disinformation or targeted harassment?

Without clear standards and oversight, tools meant to connect and inform can easily become instruments of control.

Cognitive warfare ignores national borders. State and non-state actors can influence discourse in distant countries with minimal physical presence. This makes international norms and cooperation essential: unilateral regulation in one jurisdiction can easily be bypassed via servers, shell companies, or operators elsewhere.

At the same time, heavy-handed controls risk sliding into censorship and information control of their own. The core challenge is to balance the benefits of digital technologies with strong protections for autonomy, privacy, and human rights.

Combating Cognitive Warfare

There is no single defense against cognitive warfare, but several strategies can help.

Policy and Institutional Responses

States can:

- Enact regulatory measures: Require transparency for political advertising and coordinated campaigns; impose penalties for deliberate, large-scale disinformation that causes demonstrable harm.

- Strengthen institutional trust: Improve transparency, responsiveness, and accountability so that citizens have credible alternatives to conspiracy narratives.

• Use AI defensively: Deploy detection tools to identify bot networks (botnets), deepfakes, and coordinated inauthentic behaviour, and to flag or demote clearly deceptive content.

• Cooperate internationally: Share intelligence, best practices, and technical tools to identify and disrupt cross-border influence operations.

Building Individual and Societal Resilience

Media and information literacy programs can help people recognize bias, verify sources, and resist manipulative narratives.

Studies suggest that students who receive structured media literacy education are better able to identify misleading or one-sided reporting. Taiwan, for example, has combined media-literacy curricula, independent fact-checking, rapid-response "humour over rumour" campaigns, and strengthened cybersecurity to counter misinformation linked to Chinese Communist Party influence efforts. [234]

Support for people who have been targeted—through counselling, community-building, and transparent communication from trusted leaders—can help repair the emotional and social damage caused by targeted disinformation or online harassment.

Inoculation Theory

"Inoculation theory" offers another promising approach. [235] Just as vaccines expose the immune system to a weakened version of a pathogen, this method exposes people to simplified versions of misleading arguments along with explanations of why they are wrong. Research suggests that this can "pre-bunk" certain forms of manipulation, making individuals more resistant when they encounter stronger, real-world examples later.

Games and interactive tools based on inoculation principles—such as those developed by academic and non-profit groups—show that short, engaging interventions can measurably improve people's ability to spot common manipulation techniques.

Conclusion

Cognitive warfare represents a new frontier in conflict, targeting the mind rather than physical territory. Drawing on the long history of propaganda and psychological manipulation, modern adversaries exploit digital technologies, big data, and increasingly sophisticated AI to influence behaviour, sow division, and erode trust. As these techniques grow more powerful and less visible, the ethical stakes rise.

Key questions follow: How can we preserve autonomy in an age of algorithmic manipulation? What safeguards are needed to ensure that tools meant to connect and empower us do not become instruments of coercion and control? And how should societies respond when cognitive warfare is integrated into broader strategies of state power?

In the next chapter, we turn to a concrete case: how Russia has used cognitive warfare both to shape its own population's perceptions and to undermine the stability of other nations.

CHAPTER 21 – RUSSIA'S COGNITIVE WARFARE TACTICS

Introduction

In an era when information warfare has become global, Russia stands out for the sophistication and persistence of its cognitive tactics. From the Soviet period to today's hybrid conflicts, the Kremlin has leveraged manipulation of perceptions, narratives, and emotions to achieve geopolitical goals, adapting these methods to new technologies and social vulnerabilities. This chapter examines Russia's cognitive warfare techniques, drawing on historical and contemporary examples that show their impact at home and abroad.

Control of Citizens in the USSR

Necropolitics in the Soviet Era

The Union of Soviet Socialist Republics (USSR), established in 1922 after the Russian Civil War, united 15 republics under Communist Party rule, with Russia as the dominant center. Necropolitics—the use of state power to decide who lives and who dies—was integral to Soviet control, especially under Joseph

Stalin, when millions perished through forced labor, engineered famines, deportations, and purges.

Popular uprisings and protests were met with overwhelming force. Key examples include the Hungarian Revolution of 1956, crushed by Soviet tanks; the Prague Spring of 1968, ended by Warsaw Pact invasion; long-running Baltic resistance from the 1940s to the 1980s; the Jeltoqsan protests in Kazakhstan in 1986; and the 1989 demonstrations in Georgia, all brutally suppressed. These episodes illustrate how the USSR relied on coercive violence to sustain its dominance over subject populations.

Psychological Control During the Soviet Era

Beyond physical repression, the Soviet system deployed a broad repertoire of psychological control to suppress dissent, enforce loyalty, and manufacture a unified Soviet identity.

Divide and rule

Moscow exploited ethnic, linguistic, and regional differences within and between republics to foster mistrust and dependence on the center. Arbitrary borders in Central Asia separated communities of Uzbeks, Tajiks, and Kyrgyz, entrenching rivalries. Forced deportations in the 1940s, including Crimean Tatars and Chechens, fractured social cohesion and generated enduring tensions with local populations.

Cultural suppression and Russification

Through Russification, Russian language and culture were imposed across the union. Education systems were restructured to prioritize Russian; publishing and broadcast in local languages were curtailed or censored. Writers, clergy, and intellectuals who resisted faced harassment, exile, imprisonment, or execution. Resistance to Russification in republics such as Georgia, Ukraine, and the Baltic states later fed into strong independence movements.

Intimidation and show trials

Public show trials dramatized the consequences of disloyalty. The Moscow Trials of 1936–1938 charged prominent Bolsheviks with treason on fabricated evidence; torture-extracted confessions and staged proceedings ended in executions intended to terrorize actual and potential dissenters.

Propaganda and ideological indoctrination

Ideological training saturated schools and youth organizations such as the Young Pioneers and Komsomol, glorifying Soviet leaders and portraying the party as infallible. Mass media—press, radio, and later television—operated under strict censorship, circulating only pro-Soviet narratives. Elections were carefully choreographed plebiscites that reinforced the appearance of popular consent while offering no real alternative.

Surveillance, fear, and censorship

Security agencies, culminating in the KGB, built a pervasive surveillance apparatus. Citizens lived with the knowledge that informers, phone taps, and mail monitoring could turn private grumbling into a political case. Dissidents risked prison, psychiatric incarceration, or exile. Literature, film, theater, and visual arts were tightly controlled to exclude content that might inspire critical thinking or mobilize opposition.

Manipulation of political elites

Local party leaderships were kept weak and divided. Moscow favored loyalists, rotated cadres, and encouraged rivalries among republican elites to prevent any cohesive, autonomous power base from emerging.

Together, these tools combined physical repression with deep psychological conditioning, laying a template for later Russian approaches to cognitive control.

Russian Imperatives Since the Dissolution of the USSR

On 25 December 1991, Mikhail Gorbachev resigned as president of the USSR and transferred control of the Soviet nuclear arsenal to Boris Yeltsin, president of the Russian Federation. The following day, the Soviet parliament formally dissolved the union, recognizing the independence of its 15 constituent republics. The Soviet flag was lowered over the Kremlin and replaced by the Russian tricolor.

Many influential Russians, including Vladimir Putin, later portrayed this collapse as a geopolitical catastrophe. NATO's 1999 enlargement to include Poland, Hungary, and Czechia, and NATO's intervention in Kosovo, intensified a narrative of Western encroachment. In response, the Kremlin has sought to re-assert influence over its former sphere through hybrid warfare—blending conventional forces with cyber operations, economic coercion, and cognitive warfare.

Key waypoints in this campaign include the 2008 war in Georgia, the 2014 annexation of Crimea, and the full-scale invasion of Ukraine launched in 2022. Earlier, in 2007, a massive denial-of-service attack on Estonia crippled government, banking, and media sites, demonstrating Russia's capacity to weaponize cyberspace against a neighboring state. Throughout, the Kremlin has used persistent cognitive operations to erode resistance, legitimize its actions at home, and shape international perceptions.

Russia's Cognitive Warfare Techniques

Russian strategists often frame the Soviet Union's Cold War defeat as the result of Western psychological operations—propaganda, cultural influence, and information campaigns that undermined communist legitimacy. In the post-Soviet period, Moscow has invested heavily in its own cognitive toolkit, aimed at domestic control, pressure on neighboring states, and disruption of rival democracies.

President Vladimir Putin routinely invokes imperial precedents and historical grievances, comparing himself to figures such as Peter the Great and casting contemporary invasions as the "restoration" of historically Russian lands. Recognizing the limits of its conventional military power, the Kremlin often uses information operations, subversion, and influence campaigns to soften targets and create pretexts before or alongside kinetic action.

Techniques to Control the Russian Populace

Surveillance and data-driven control

Inside Russia, the state has expanded a dense surveillance infrastructure. Moscow alone has deployed a vast network of facial-recognition-enabled cameras, and similar systems are spreading to other cities. These tools are used not only for routine policing but also to identify draft evaders, protesters, and political activists. The System for Operative Investigative Activities (SORM) gives the Federal Security Service (FSB) direct access to phone and internet traffic, enabling large-scale interception and storage of communications. Complementary systems such as Semantic Archive mine open-source materials—blogs, social media, and online forums—to map networks and track narratives, while internet filtering blocks foreign sites and forces removal of domestic content judged "extremist" or disloyal.

Propaganda and "othering"

Domestic messaging paints Russia as a besieged fortress surrounded by decadent, hostile "others"—often labeled "Nazis," "Russophobes," or agents of Western moral decay. Necropolitical rhetoric glorifies sacrifice and violence against designated enemies as morally necessary. Government-Organized Non-Governmental Organizations (GONGOs) echo state narratives under the appearance of civic initiative, while the Russian Orthodox Church frequently sacralizes Kremlin policies, framing geopolitical conflicts as spiritual battles.

Patriotic education and militarization of youth

Since the full-scale invasion of Ukraine, budget outlays for "patriotic education" have risen dramatically, funding youth movements, school rituals, and curricula that emphasize obedience, militarism, and national greatness. Children participate in flag-raising ceremonies, anthem singing, and weekly civics lessons that justify the war and demonize perceived enemies. These programs seek to embed state narratives early and normalize permanent mobilization.

Media manipulation and controlled narratives

State-aligned television, news agencies, and online portals constantly stress external threats—from NATO, Ukraine, and the "collective West"—while presenting the Kremlin as a protective, benevolent authority. Disinformation campaigns, such as inflated claims of persecution of Russian speakers in Ukraine, have been crucial in justifying interventions like the annexation of Crimea and the invasion of Donbas. Independent outlets face legal harassment, blocking, or branding as "extremist."

Crushing dissent and stigmatizing "foreign agents"

Successive "foreign agent" and "undesirable organization" laws have targeted NGOs, media, and individuals receiving any foreign support or publishing critical content. Designation brings onerous reporting requirements, stigmatizing labels, and often effective closure. High-profile opposition figures—including Alexei Navalny—have faced imprisonment, attempted assassinations, and ultimately death, sending a clear signal about the personal risks of contesting the regime.

Russia's Ongoing External Cognitive Warfare Campaign

Beyond its borders, Russia has refined cognitive warfare tactics to target foreign populations, political systems, and alliances, increasingly leveraging AI and digital platforms to increase scale and precision.

Intimidation and coercive signaling

Russian forces routinely conduct military maneuvers near NATO borders and stage airspace incursions to project menace and test responses. In 2013, long-range bombers carried out a simulated attack profile against targets in Sweden, approaching within tens of kilometers of Swedish territory. Subsequent years have seen numerous airspace violations over the Baltic region and aggressive maneuvers near allied aircraft and ships, reinforcing a narrative of Russian reach and Western vulnerability.

Historical revisionism and narrative warfare

Kremlin messaging foregrounds Russia's role in defeating Nazi Germany while downplaying or justifying episodes such as the Molotov–Ribbentrop Pact and post-war repression in Eastern Europe. Official statements and state media portray contemporary neighbors—especially Ukraine and Baltic states— as "artificial" or "historically Russian" territories. These narratives are used to legitimize territorial grabs and to cast opposition to Russian expansion as neo-Nazi or fascist.

Propaganda and information laundering

State-funded outlets such as RT and Sputnik package Kremlin positions as alternative journalism for global audiences. They tailor content to local grievances—anti-elitism, anti-Americanism, Euroscepticism—to build trust while systematically amplifying pro-Russian frames. These channels often launder narratives that originate in anonymous blogs, fringe sites, or intelligence-linked platforms, re-circulating them across multiple layers until their origins are obscured.

Disinformation campaigns against foreign societies

Russian information operations routinely spread false or distorted stories aimed at inflaming social and cultural conflicts. Examples include claims that Sweden "kidnaps" Muslim children into state care, or that Finnish authorities systematically remove children from immigrant families—narratives that have sparked diplomatic friction and protests. Such campaigns seek to erode trust in state institutions, heighten minority-majority tensions, and complicate Western policy toward Russia.

Symbolic violence and demonstrations of force

Highly publicized military exercises, snap drills, and nuclear signaling serve as symbolic violence directed at external audiences. These spectacles are designed not only to train forces but to reinforce an image of Russia as unpredictable and willing to escalate, influencing foreign threat perceptions and domestic debates about defense spending and alliance cohesion.

Social media manipulation, troll farms, and bots

Since at least 2014, Russia's Internet Research Agency and similar entities have run troll farms and bot networks to interfere in foreign politics. During the 2016 US presidential election, Russian-linked accounts on Facebook, Instagram, and Twitter reached tens of millions of users with divisive content about race, migration, guns, and LGBTQ rights, often masquerading as grassroots activists. Coordinated "computational propaganda" campaigns have also targeted the Brexit referendum and elections across Europe, Africa, and Latin America, exploiting each society's specific cultural fault lines.

Deepfake technology and AI-generated personas

Russian-linked actors have experimented with deepfake videos, synthetic news anchors, and AI-generated personas to degrade trust in visual evidence and to push Kremlin-aligned narratives. AI-produced media allows rapid, low-cost creation of "local" influencers, fake experts, and tailored propaganda, blurring the line between authentic and fabricated discourse and making detection more difficult.

Implications and Ethical Concerns

Russia's cognitive warfare tactics raise far-reaching ethical and legal concerns. By weaponizing information and systematically sowing mistrust, these operations corrode confidence in institutions, news media, and even intimate relationships, leaving societies polarized and easier to manipulate. Interference in elections and referendums threatens democratic self-determination and can delegitimize outcomes for years.

The integration of AI, big-data analytics, and pervasive surveillance into propaganda systems requires extensive collection and processing of personal data, often without consent, posing grave threats to privacy and autonomy. The covert, deniable nature of these operations makes them hard to detect, attribute, and regulate, complicating efforts to hold perpetrators accountable and to design proportionate countermeasures that do not themselves erode civil liberties.

Conclusion

Russia's cognitive warfare practices, grounded in Soviet traditions of repression and propaganda but augmented by contemporary technologies, constitute a serious challenge to global security and democratic resilience. By combining old-style state control with big-data profiling, social-media manipulation, and AI-driven disinformation, the Kremlin continues to shape narratives, suppress dissent, and project power far beyond its borders. Responding effectively will require sustained international cooperation, careful norm-building, and vigilant domestic policies that strengthen societal resilience without replicating the very authoritarian methods they seek to resist.

CHAPTER 22 – AI-ENABLED COGNITIVE WARFARE

Introduction

The tools of mind control have evolved from crude propaganda and terror into something far more pervasive and personalized. Traditional techniques—state propaganda, "othering" of enemies, and exploitation of group dynamics—have been fused with big-data analytics, generative AI, and neurotechnology to create modern cognitive warfare: the deliberate targeting of how people think, feel, and decide, at scale, in peace and war. This chapter traces that evolution and examines how major powers now weaponize AI-enabled cognitive tools as part of hybrid warfare in conflicts from Ukraine to Gaza and the wider confrontation with Iran.

From Propaganda and "Othering" to Cognitive Warfare

Early forms of mind control relied on centralized propaganda, censorship, and coercion. Totalitarian regimes used one-to-many media to push simple narratives: glorification of the leader, demonization of internal and external enemies, and manufactured consensus through orchestrated elections and show trials. These approaches leveraged fundamental aspects of group behaviour, including conformity, adherence to authority, loyalty within groups, and apprehension regarding exclusion or disciplinary action. Cognitive warfare retains these foundations but adds new layers. Contemporary analyses describe it as "an emergent form of warfare" that goes beyond classic psychological operations by combining disinformation, emotional manipulation, and exploitation of cognitive biases (such as confirmation bias and anchoring) with strategic timing and advanced technologies. [236] In hybrid warfare, such campaigns are integrated with cyberattacks, economic pressure, and kinetic force to paralyze decision-making, erode social cohesion, and reshape perceptions of legitimacy.

AI-driven information warfare represents a further step-change: machine-learning systems can ingest vast amounts of behavioural data, segment audiences, and generate tailored messages that "spot vulnerabilities within target groups and tailor their influence efforts to control public opinion, shape political debates, and even disrupt democratic processes." [237] Deepfakes and AI-generated news blur the boundary between reality and fabrication, creating what one report calls a "crisis of knowing."

AI-Enabled Techniques in Contemporary Hybrid Wars

Russia and Ukraine

Russia's operations around the 2022 invasion of Ukraine are widely described as a laboratory for cognitive warfare. A primer on Russian cognitive warfare argues that Moscow uses cognitive tools "to facilitate its war in Ukraine, shape Western decision-making, obfuscate objectives, preserve Putin's regime, and mask Russia's weaknesses," making the human mind "a central battlespace."[238] A *Frontiers in AI* case study of the conflict shows how manipulative narratives—especially about nuclear escalation—were timed to exploit fear, information overload, and confirmation bias in Western audiences, amplifying calls for restraint and undermining support for Ukraine.[239]

Recent work documents AI-driven disinformation campaigns on Twitter/X in the war, with Russian-linked operators using machine-learning tools to analyze sentiment and then push tailored propaganda that depicts Ukraine as corrupt or doomed and questions the value of Western aid.[240] At the same time, deepfake videos—such as fabricated clips of Ukrainian leaders calling for surrender—illustrate how AI-generated media are being used to sow confusion, erode trust in visual evidence, and test the boundaries of plausibility.

Ukraine has not remained passive. Studies of "Ukraine and the future of cognitive warfare" highlight how Kyiv has developed a whole-of-society information defense, combining strategic communications with volunteer OSINT networks and AI-supported verification tools.[241] Ukrainian projects have used facial-recognition and open-source data to identify dead Russian soldiers and message their families, seeking to puncture Kremlin narratives and create cognitive dissonance inside Russia itself.[242] In this sense, both sides are using evolved propaganda, othering, and group psychology—now supercharged by AI—as core components of hybrid war.

China's doctrine goes further in formalizing AI as a tool of cognitive warfare. PLA writings on "cognitive domain operations" describe using big-data and AI systems to profile target groups, identify emotional and ideological vulnerabilities, and deliver "spiritual ammunition" (information payloads) through social media and digital platforms to induce "internal psychological conflicts."[243]

Analyses of Chinese cognitive warfare emphasize the integration of public-opinion warfare, psychological warfare, and legal warfare, with AI used to automate sentiment analysis, generate tailored content, and algorithmically prioritize pro-CCP narratives, including via apps such as TikTok/Douyin.

US doctrine, by contrast, is more ambivalent. A US Special Operations Command procurement document leaked to the press reveals plans for "a next generation of 'deep fake' or other similar technology to generate messages and influence operations via non-traditional channels," including life-like fake personas and videos for military information support operations.[244] Analysts note that such tenders show active experimentation with offensive deepfake psy-ops, even as official statements warn that the Pentagon "should not be fighting fire with fire" in disinformation. Public US and NATO strategies emphasise AI mainly for detection and resilience against adversaries' cognitive warfare, but the leaked documents demonstrate that Western militaries are also exploring how these tools might be weaponized.

Recent conflicts in the Middle East show Israel and Iran deploying similar cognitive-warfare techniques. Analyses of the 2023 Israel–Hamas war describe a "cognitive battlefield" in which both sides used bots, coordinated campaigns, and AI-generated imagery to control global narratives and influence diasporas.[245]

Investigative reports reveal that Israel's Ministry of Diaspora Affairs covertly funded the "Stoic" operation, which used fake social-media accounts and AI-generated content to target US and other legislators with pro-Israel messaging about the Gaza war.[246] Commentators describe this as a clear example of AI-assisted foreign influence built on classic propaganda and lobbying techniques.

Iran, for its part, has been expanding its digital frontlines. A Middle East Institute study of a 12-day Israel–Iran confrontation notes that Iranian operators used coordinated networks of accounts and AI-generated disinformation to flood social media with anti-Israel and anti-protest messaging, including impersonating Israelis online.[247]

Other analyses show Iran-linked influence operations targeting Western debates on Iranian uprisings, using botnets and inauthentic personas to discredit protesters and amplify regime narratives.[248]

These campaigns rely on familiar "othering"—portraying opponents as terrorists, foreign agents, or immoral degenerates—now delivered via AI-assisted account management and content generation.

European and NATO responses show how defensive cognitive tools are being built on the same technological foundations. EU-funded projects in the "AI against disinformation" cluster develop machine-learning systems to detect coordinated inauthentic behaviour, flag deepfakes, and support fact-checkers, explicitly framed as counters to Russian and Chinese cognitive warfare rather than tools for offensive propaganda. NATO experimentation with deepfakes in wargames has revealed how fabricated news videos can mislead even experienced officers, leading the alliance to invest heavily in detection and cognitive resilience exercises.[249] These developments underscore that cognitive warfare has become a central concern for democratic states, even if their declared emphasis is still on defense.

Leaked Experiments and the Normalization of Deepfakes

Deepfakes epitomize the convergence of old propaganda with new AI. They enable the creation of "highly convincing fake videos and audio recordings that can be used to manipulate public opinion, spread false information, and incite political unrest," and have already been used to fabricate statements by political leaders in multiple conflicts. Russian and pro-Russian actors have tested deepfakes of Ukrainian President Volodymyr Zelenskyy and other officials, while Chinese and pro-PRC networks have experimented with AI-generated news anchors and synthetic personas.

Leaked and procurement documents show that Western militaries are not only devising ways to detect deepfakes but also considering their offensive use. SOCOM's wish-lists for "deep fake" capabilities, along with revelations that the US military conducted online influence operations using fake news sources and AI-generated avatars on platforms such as Twitter, indicate that deepfakes are being treated as potential tools for Military Information Support Operations.[250] At the same time, defense agencies fund research into AI-powered deepfake detection and issue public warnings about the dangers of deepfake-enabled cognitive warfare, reflecting the tension between deterrence, reciprocity, and normative restraint.

The Next Frontier: Brain–Computer Interfaces and Neuro-Enabled Warfare

Beyond information ecosystems, advances in brain–computer interfaces (BCIs) suggest a coming fusion of neurotechnology and cognitive warfare. DARPA programs such as the Next-Generation Nonsurgical Neurotechnology (N3) project aim to develop non-invasive neural interfaces that would allow operators to control swarms of drones, cyber-defense systems, or other assets "at the speed of thought."[251] In trials, a paralyzed volunteer used a bidirectional neural implant and EEG-mediated interface to steer simulated unmanned aerial vehicles and receive sensory feedback directly to his brain, illustrating the feasibility of thought-controlled systems.[252] DARPA officials have sketched a future in which service members can put on non-invasive neural interface headgear easily as other tactical gear and, for the duration of a mission, use it to interact with unmanned vehicles, cyber-defense systems, and decision-support software before taking it off again.

Ethicists warn that as AI-augmented BCIs move from medical rehabilitation into military enhancement, they raise profound concerns about cognitive liberty, mental privacy, and autonomy. A review of AI-assisted warfighter enhancement notes that implantable or wearable BCIs could enable "bionic super-soldiers" but also expose them to hacking, unauthorized data extraction, and unprecedented command-and-control over their mental states.[253] In the longer term, integrating BCIs into command structures could allow for more seamless team coordination—such as networked troops sharing EEG-derived intent signals—but also introduces the possibility of direct neuromodulation as a tool of discipline or coercion.

From a mind-control perspective, BCIs and neuro-enhancement do not replace cognitive warfare; they fold the nervous system itself into the battlespace, turning neural signals into both inputs and outputs for military systems. Combined with AI-enabled targeting and deepfake-driven information environments, they point toward a future in which the line between psychological operations and direct neuro-intervention becomes increasingly blurred.

Conclusion

The trajectory from traditional propaganda and othering to AI-driven cognitive warfare and neuro-enabled soldier enhancement illustrates how mind-control techniques are being systematically weaponized in modern conflict. States such as Russia, China, the US, the UK, Ukraine, Israel, and Iran now treat the cognitive domain as a central theatre of operations, leveraging AI, big data, and, increasingly, neurotechnology to shape perceptions, emotions, and decisions in ways that would have been unimaginable in the 20th century.

This chapter completes the picture of how minds are targeted in contemporary war. The chapters that follow turn to the remaining questions: how othering and dehumanization can be countered in democratic societies, and what ethical and legal constraints should govern the use of AI and neurotechnology in order to protect cognitive autonomy and human dignity in an age of weaponized mind control.

CHAPTER 23 – OVERCOMING "OTHERING"

Introduction

History shows that human beings are remarkably vulnerable to psychological manipulation, particularly through the divisive tactic of othering. From early propaganda and ritualized exclusion to modern digital and neurotechnological tools, manipulative actors have exploited perceived differences to divide, weaken, and control societies. In today's digital environment, the stakes are higher than ever, making the protection of psychological autonomy a central ethical concern.

Recognizing the scale and seriousness of psychological manipulation is the essential first step in limiting its destructive power. Episodes ranging from the propaganda of totalitarian regimes and the crimes of Unit 731 to covert programs such as MKUltra and recent data-driven operations like Cambridge Analytica demonstrate how easily "difference" can be weaponized. Denying or downplaying these patterns only increases societal vulnerability. This chapter is therefore a call to action, urging international organization's, states, communities, and individuals to confront "othering," protect cognitive autonomy, and uphold ethical standards as the basis for a more resilient and inclusive future.

Combating "Othering" at the International Level

Global cooperation is pivotal in addressing the widespread and multifaceted nature of psychological manipulation. Shared norms and practical frameworks can help counter divisive tactics that operate both online and offline.

Voluntary Service Overseas (VSO)[254]

Voluntary Service Overseas (VSO), based in London, is an international development organization that works with communities in some of the world's poorest regions. Recognizing that "othering" can undermine its mission, VSO has introduced inclusivity training that emphasizes:

- Acknowledging and challenging personal biases: Reflecting on one's own prejudices as a starting point for dismantling internalized barriers.

- Cultural open-mindedness: Practicing cultural humility by rejecting assumptions of superiority and valuing diverse perspectives.

- Mindful language: Choosing words that foster inclusion rather than reinforce stereotypes.

- Recognizing intersectionality: Understanding that identities are multifaceted and overlapping.

- Speaking up: Actively challenging biased behaviour so that "othering" becomes socially unacceptable rather than silently tolerated.

Based in Belfast, the Social Change Initiative (SCI) advances social justice, human rights, and peacebuilding through advocacy and community engagement. Its work against "othering" includes:

- Global networking: Linking activists and organizations to create mutual support and shared learning.

- Strategic advocacy and support: Helping partners design messaging and communications to address hate and extremism.

- Research and documentation: Gathering evidence from campaigns worldwide to identify effective methods.

- Learning exchanges and study visits: Bringing together activists and community leaders from different countries—for example, a delegation to Greece to examine responses to the far-right Golden Dawn.

- Emphasis on intersectionality: Designing programs that address how overlapping identities shape experiences of discrimination.

- Creation of safe spaces: Encouraging environments where people can discuss experiences, reflect on bias, and build solidarity without fear.

The Democracy and Belonging Forum[256]

Hosted by the Othering & Belonging Institute at UC Berkeley, the Democracy and Belonging Forum focuses on the roots of division and exclusion by:

- Bridging divides: Connecting civic leaders across Europe and North America to address political, ideological, and identity-based polarization.

- Conducting research and offering resources: Analyzing authoritarian populism and fragmentation, and providing tools to counter them.

- Building coalitions: Encouraging diverse groups to collaborate in forming a broader, more inclusive "we."

- Advancing belonging: Working to ensure that historically marginalized voices are fully included in democratic processes and institutions.

The United Nations Development Programme (UNDP)[257]

Active in more than 170 countries and territories, the UNDP works to eradicate poverty, reduce inequality, and build resilience to crises. Recognizing that misinformation, disinformation, and hate speech can fuel "othering," it has adopted several innovative tools:

- The iVerify platform: A system combining AI and human fact-checking to assess the veracity of information, especially around elections, and flag false or misleading narratives.

- Accelerator Labs: Initiatives that map and evaluate efforts to counter misinformation, particularly in the Global South, and propose improvements in international coordination and partnership building.

Combating "Othering": Individual Responsibility and Empowerment

Institutions matter, but every individual also plays a crucial role in countering "othering" and protecting their own psychological integrity. Strengthening critical thinking, media literacy, emotional intelligence, and empathy enables people to recognize and resist divisive narratives before they take root. Acknowledging one's own biases and emotional triggers is a necessary starting point.

Beyond personal reflection, individuals can demand transparency and accountability from institutions that shape public discourse—governments, technology companies, media organizations, and political parties. Participation in civic life through voting, public consultation, policy advocacy, and community organizing helps reinforce democratic norms and builds collective resilience against manipulation.

Education's Role in Countering Propaganda and Misinformation

Education remains one of the most powerful tools against psychological manipulation. School curricula need to evolve to address the challenges of digital propaganda and organized disinformation. A robust approach to media literacy should teach students to analyze, evaluate, and create media messages across platforms. Done well, such education not only builds resistance to manipulative narratives but also fosters empathy by encouraging students to examine bias—both in media and in themselves. Students can, for example, learn to:

- Identify stereotypes and recognize common propaganda techniques.

- Analyze how media messages are constructed and what purposes they serve.

- Reflect on their own media consumption and how it influences their beliefs.

Many countries have begun integrating media literacy into their education systems:

- Mandatory courses: Finland, Denmark, France, South Korea, and Uganda have introduced dedicated media literacy requirements.

- United States initiatives: States such as Florida and Ohio have incorporated media literacy into K–12 curricula, with additional states developing similar programs.

- Integrated subjects: In the UK, Canada, Australia, and elsewhere, media-literacy components are embedded in subjects like English, Citizenship, and Social Studies, even where they are not stand-alone courses.

It is important to distinguish between media literacy (the ability to critically evaluate media) and media information (content and distribution channels). For example, New Jersey's 2023 mandate on "information literacy" in K–12 education introduces valuable skills for evaluating sources, but does not yet extend to more advanced issues such as cognitive warfare techniques, psychological targeting, or state-sponsored disinformation. Broadening such programs to cover these areas remains an outstanding challenge.

Public Awareness Campaigns

Public awareness campaigns that highlight the dangers of "othering," propaganda, and misinformation can further strengthen societal resilience. Regular, honest communication about these risks helps communities to recognize patterns of manipulation and respond proactively. Transparent dialogue encourages a shared commitment to ethical communication and to the protection of collective autonomy.

Several countries have already launched initiatives of this kind. Germany's No Hate Speech Movement, initiated by the Council of Europe in 2013, mobilizes young people, educators, and activists to challenge hate speech and promote human rights online.[258] Through workshops, educational materials, and coordinated campaigns, it offers practical strategies for contesting hateful narratives and amplifying inclusive ones.

Similarly, the Australian Human Rights Commission's "Racism. It Stops With Me" campaign, launched in 2012, addresses discrimination and promotes social cohesion, thereby indirectly confronting "othering."[259] Working with schools, businesses, and community organizations, the campaign uses multimedia messaging to prompt reflection on biases and behaviours, and provides tools for building more inclusive workplaces and communities. Its Workplace Cultural Diversity Tool, for example, helps organizations initiate constructive discussions about cultural diversity and anti-racism, including issues affecting First Nations peoples.

While these initiatives have clearly increased awareness and mobilized communities, rigorous evidence that they reduce hate speech or discrimination at a national level remains limited. Measuring such causal effects is difficult: data collection methods vary, hate speech often moves across platforms and private channels, and trends are influenced by parallel factors such as legislative changes, economic conditions, and broader social movements. Nonetheless, these campaigns provide important models for how states and civil society can work together to counteract "othering."

Conclusion

Psychological manipulation and "othering" have marked both past and present, but they are not inevitable. Humanity's capacity for empathy, solidarity, and ethical self-restraint offers a powerful counterweight. By empowering individuals through education, fostering communities grounded in mutual respect, sustaining institutional integrity, and building international cooperation, societies can weaken the hold of "othering" and move toward a future defined by dignity and shared belonging.

At the same time, a larger challenge is emerging. Just as many societies are learning to push back against misinformation, hate speech, and overt forms of dehumanization, the rapid growth of artificial intelligence and neurotechnologies threatens to amplify these dangers. Without clear constraints, these tools could undermine autonomy, exacerbate division, and erode social cohesion. The following chapters turns to this frontier, examining how AI and neurotechnology must be governed if we are to protect cognitive freedom in an era of weaponized mind control.

CHAPTER 24 – THE MUSK FACTOR

Introduction

For most of the twentieth century, the idea of mind control belonged to the domains of secretive state programs, covert military experiments, and fringe cults. In the twenty-first, it has begun migrating into glossy product launches, investor presentations, and social-media hype. No single figure embodies this shift more clearly than Elon Musk. Through a sprawling empire that includes electric vehicles, space launch and satellite networks, social media, brain–computer interfaces, and humanoid robots, Musk has positioned himself not only as an industrialist, but as an architect of the environments in which people live, communicate, and increasingly, think.

My own decision to write this book was triggered by one part of that empire: Neuralink, Musk's brain-implant company. Watching invasive neurotechnology presented as a lifestyle upgrade—with little public discussion of its potential for manipulation, surveillance, or coercive control—made one thing clear. Mind control is no longer a historical horror or a speculative threat. It is entering the commercial mainstream, driven by entrepreneurs who answer primarily to markets and to their own ambitions.

This chapter examines "the Musk factor" not because Musk is uniquely evil, but because he is uniquely concentrated: a single individual whose companies reach from orbit to the human cortex.

He illustrates what happens when technological power over perception, communication, and the brain accumulates in private hands with few effective constraints.

Musk's Psychology and Public Persona

Musk's rise has been built on intelligence, vision, relentless work, and a willingness to take large risks with other people's money and lives. It has also been marked by volatility and controversy. In 2017, in an exchange on what was then Twitter, he described his own experience as "great highs, terrible lows and unrelenting stress," and loosely agreed with a follower's suggestion that he might be bipolar—before later hedging that he was not "medically" diagnosed.[260] At other times, he has hinted at depression and publicly defended his use of ketamine as a treatment, arguing that for some individuals it may be preferable to conventional antidepressants.

None of this is inherently disqualifying; many capable leaders live with mental-health challenges. But when a person with pronounced mood swings, impulsive posting habits, and a taste for public feuds gains influence over tools that can reshape information flows and directly interface with the brain, their psychological profile becomes ethically relevant. It shapes how they respond to criticism, handle risk, and interpret the idea of "acceptable collateral damage."

Musk's personal life reflects the same pattern of intensity and instability. He has fathered fourteen children with four different partners, while maintaining a public persona that oscillates between self-styled savior of civilization and aggrieved victim of his own success. The goal here is not to police his private relationships, but to note the consistent theme: a man who spreads himself widely across projects, people, and promises, yet demands loyalty and control wherever his attention lands.

Inside the Musk Workplace

Accounts from former employees portray Musk's companies as simultaneously exhilarating and brutal. A Los Angeles *Times* article from November 2022 summarized his management style this way[261]:

> If there's such a thing as a warm and cuddly boss, Musk has long been the opposite to his employees, who now number more than 100,000. He burns through executives with the heat of a battery fire. He takes criticism personally, even when it's a matter of worker or customer safety. He' has been known to fire people on a whim.

Stories from Tesla, SpaceX, and later Twitter/X all follow similar lines: sudden firings, public humiliation of staff, impossible deadlines, hostility to internal dissent, and a tendency to override engineers and safety experts when they raise concerns.[262] Within limits, such behaviour can be framed as the archetypal "hard-driving founder" trope. Beyond those limits, it becomes an organisational pathology: a culture where the boss's mood outranks evidence, and where people who question risk are treated as traitors rather than guardians.

That pattern matters profoundly when moving from electric cars or rockets—which are dangerous enough—to implants in human brains. A leader who treats caution as disloyalty and regulatory scrutiny as personal insult is poorly suited to stewarding technologies where a single design flaw can damage cognition or personality.

Political Power and the Capture of the State

Over the last two decades, Musk's companies have received tens of billions of dollars in public funding, contracts, and subsidies: from NASA launch contracts and Pentagon satellite deals to tax credits for electric vehicles and renewable energy. For years, he sold himself as a politically eclectic innovator whose work transcended partisan divides. That posture shifted sharply as he began to position himself as a tribune of the "anti-woke" right.

By mid-2024 Musk had formally endorsed Donald Trump's return to the presidency and poured more than two hundred million dollars into supporting his campaign and aligned political action committees.[263] In return, Trump lavished praise on Musk as a visionary job-creator and promised to cut "red tape" that allegedly held American innovation back. When Trump returned to office in January 2025, Musk's influence moved from rhetorical to structural.

First, Musk's social-media platform, now rebranded as X, became an unofficial megaphone for the new administration's messaging. He reinstated previously banned far-right accounts, amplified conspiracy-driven narratives, and harassed journalists who criticized his companies or his politics. Progressive voices and independent researchers found themselves throttled, dogpiled, or suspended, even as extremist influencers were granted direct access to Musk and his team.

Second, Trump tasked Musk with leading an effort to "rationalize" federal agencies: a euphemism for firing tens of thousands of civil servants, many of them in regulatory or scientific roles. Among the hardest hit were bodies responsible for environmental protection, labour standards, and the oversight of advanced technologies. In effect, Musk helped design the state that would then be asked to regulate his own interests.

It is in this context that Neuralink must be understood.

Neuralink: Colonizing the Cortex

Neuralink, founded in 2016, is Musk's bid to link human brains directly to computers. Its devices are coin-sized implants that sit in the skull, with ultra-fine threads penetrating the cerebral cortex to record neural activity and, eventually, stimulate it. The company markets its work as a way to restore movement and communication to people with paralysis, and there is genuine promise in that mission. But Musk's rhetoric quickly leaps beyond therapy. He has repeatedly framed Neuralink as a path for ordinary people to "keep up" with artificial intelligence, implying widespread adoption of brain implants in healthy individuals as a kind of arms race against the machines.[264]

Even in purely medical use, the risks are substantial. Surgical implantation carries dangers of bleeding, infection, and scarring. The threads can move or break. Batteries can fail. Removing or upgrading implants without harming surrounding tissue is difficult. Ethical critics have also warned of subtler harms: dependency on corporate-controlled hardware to perform basic functions, the possibility of personality changes from stimulation, and a blurring of the boundary between voluntary thought and device-mediated action.

These concerns are not hypothetical. Neuralink's first human trial, announced with great fanfare, soon encountered serious problems when some of the implant's threads began retracting from the brain tissue, degrading performance and forcing engineers to compensate with software.[265] The company presented this as a manageable setback, but it illustrates a basic point: we do not yet know how these devices behave over years inside a living brain. Early complications may be the rule, not the exception.

Animal-welfare investigations have added another layer of alarm. Under pressure to meet Musk's aggressive timelines, Neuralink reportedly conducted rushed experiments that led to unnecessary suffering and deaths in monkeys, pigs, and other animals.[266] Whistleblowers described repeated surgeries, inadequate planning, and a culture that prized speed over care. Federal authorities opened inquiries into possible violations of animal-welfare laws. When a company cuts ethical corners in animal research, there is little reason to assume it will be scrupulous when human subjects are at stake.

Dismantling the Referee: FDA Firings and Safety Oversight

Against this backdrop, the new administration's assault on the civil service took on a darker significance. Among the agencies hit was the US Food and Drug Administration, including its Office of Neurological and Physical Medicine Devices—the unit that reviews brain implants and similar technologies. Around twenty scientists and medical officers in that office were abruptly dismissed as part of a wider purge, several of them involved in assessing Neuralink and other brain-computer interface trials, raising concerns about the loss of expertise and independent oversight.[267]

Officials framed the firings as part of a broader "efficiency drive," insisting that remaining staff and outside contractors could handle the workload. Yet within weeks, reports emerged that the agency was scrambling to rehire experienced reviewers and that internal capacity to scrutinize complex neuromodulation devices had been seriously weakened. In other words, just as a politically connected company with a checkered safety record moved into first-in-human trials, the public's primary line of technical defense was thinned out.

Whether or not the purge was engineered with Neuralink specifically in mind, its effect was to tilt the playing field in favour of rapid approval and against cautious examination. At minimum, it sends a chilling message to regulators: push too hard on politically favored companies, and your job may disappear.

Starlink, X, Optimus: Building a Control Stack

Neuralink is only one part of Musk's emerging "control stack." Starlink, his satellite-internet constellation, already provides connectivity in remote regions, disaster zones, and war theatres. In Ukraine, Musk has alternately enabled and restricted Starlink access, affecting military operations and diplomatic strategies.[268] The decision to maintain or withdraw coverage in a given area can profoundly shape events on the ground—yet it is taken by a private executive rather than an accountable public body.

X, meanwhile, functions as a global propaganda organ tailored to its owner's preferences. Algorithmic changes that boost Musk's posts and amplify favored factions while suppressing critics effectively allow him to steer public discourse.[269] [270] Decisions about which accounts to restore or ban, which narratives to promote, and which whistleblowers to dogpile have deep consequences for democratic deliberation.

Tesla's Optimus humanoid robot extends this influence into physical space. Optimus is marketed as a general-purpose laborer and assistant, capable of performing factory work, household tasks, and potentially security roles. A fleet of networked humanoids, equipped with cameras, microphones, and sensors, integrated with Starlink for connectivity, X for data streams, and Neuralink-responsive interfaces, could become a ubiquitous, mobile layer of surveillance and enforcement. Even if Optimus never reaches such sophistication, the very idea that a single corporation might supply the robots in our factories, homes, and streets should give pause.

Add these elements together and a disturbing picture emerges. One empire aspires to control the channels through which we receive information (X), the infrastructure that carries that information (Starlink), the vehicles that move us (Tesla), the robots that work alongside or for us (Optimus), and the devices that may one day sit inside our skulls (Neuralink). In such a world, "mind control" is not a metaphor. It is an emergent property of an integrated system.

Why the Musk Factor Matters

The danger here is not simply that Elon Musk is flawed, mercurial, or politically reactionary. Many powerful people share those traits. The danger is the combination of three elements:

1. Concentration of power: One individual and his companies sit astride multiple critical infrastructures, from orbit to the cortex.

2. Erosion of checks and balances: Regulatory agencies and public institutions that might restrain excess are being weakened, captured, or threatened into compliance.

3. Technologies that penetrate the mind: Neuralink-style implants, coupled with pervasive data

collection and automated persuasion, open direct channels into thought, emotion, and behaviour.

Conclusion

Historically, mind control required crude methods: isolation, drugs, torture, relentless propaganda. Today, a handful of highly capitalized firms are building tools capable of influencing perception and cognition at scale, and in some cases, quite literally, of reading and writing neural activity. When those firms are dominated by a single, ideologically driven personality who treats dissent as treachery and regulation as an obstacle, the risk ceases to be theoretical.

This is why the Musk factor deserves its own chapter. It crystalizes the central concern of this book: that the power to shape minds, once the province of states and sects, is migrating into private empires animated by profit, ego, and political vendettas. Neuralink is not just another medical device start-up. It is a gateway through which the last frontier of human autonomy—the brain itself—may be opened to systematic external control.

The task for the following chapter is to step back from the individual and ask: what ethical frameworks, legal safeguards, and international agreements are needed to ensure that no one—whether Elon Musk or his successors—can unilaterally seize that power?

CHAPTER 25 – THE ETHICAL APPROACH TO INNOVATION

Introduction

From ancient mind control practices to cutting-edge advances in artificial intelligence and neurotechnology, one critical question persists: how can we protect the human mind and preserve individual autonomy in the face of technological power? This final chapter surveys the ethical challenges posed by both historical methods of mind control and contemporary digital and neural tools, and argues for robust ethical frameworks, international cooperation, and vigilant oversight to ensure that progress remains aligned with human dignity. The regulation of AI developments is considered first, followed by the regulation of neurotechnological advances.

The Legacy of Mind Control

Historical lessons

The history of mind control, marked by invasive psychosurgeries and dehumanizing state-run experiments, shows how the drive to control the human mind can lead to profound harms. These episodes serve as stark reminders that scientific and technological advances, however promising for treatment or enhancement, must always be weighed against their potential for misuse. Transparency, informed consent, and meaningful accountability are fundamental when intervening in the human psyche.

The dual-edged nature of innovation

Mind control technologies—whether chemical, surgical, digital, or algorithmic—have always been double-edged. They offer hope for treating otherwise intractable conditions and for augmenting cognitive abilities, yet they also risk exploitation to manipulate behaviour, suppress dissent, and centralize power. The ethical challenge is to harness these tools for genuine therapeutic and social benefit while safeguarding the very freedoms they might undermine.

Controlling AI: innovation with responsibility

The Power and Peril of Artificial Intelligence

Artificial intelligence now operates at scales where it can process vast datasets and influence opinions and decisions—both subtly and overtly. It promises breakthroughs in healthcare, education, logistics, and finance, but its capacity to shape information flows raises significant ethical concerns. Algorithm-driven social media platforms can deepen polarization by creating echo chambers, while advanced language models can generate highly persuasive yet potentially misleading or fabricated content.

The Digital Panopticon and Surveillance States

The old concept of the panopticon—a structure in which people behave as if they are under constant observation—has been reborn in digital form. Today, governments and corporations can monitor much of human behaviour through sensors, cameras, and data-mining algorithms.

In China, for example, the Social Credit System seeks to enforce state-approved behaviours through continuous monitoring and a mix of rewards and sanctions. Globally, smartphone apps, wearable devices, and smart home technologies collect detailed data on locations, habits, and biometric signals, encouraging self-censorship and quiet conformity out of fear that one is always being watched.

The ethical implications are profound. Pervasive surveillance undermines privacy, erodes trust, and pressures individuals to align with dominant norms. Such environments become fertile ground for manipulation and control, threatening free expression and personal autonomy.

Ranking Countries on Ethical Ai Governance

Analytical tools now attempt to rank countries not only on their AI capabilities but also on how responsibly they govern them. Stanford's Global AI Vibrancy Tool, for instance, continues to find the United States among the global leaders in overall AI "vibrancy," alongside China, the United Kingdom, India, and other major economies.[271] By contrast, the Global Index on Responsible AI (GIRAI) places the Netherlands first for responsible AI practices, followed by countries such as Germany, Ireland, and the United Kingdom, with the United States somewhat lower despite its technical leadership.[272]

Changes in US policy direction—particularly any reduction in safeguards or oversight—may weaken its standing in such indices over time and complicate cooperation with jurisdictions that prioritize strong rights-based governance.

The Need for Ethical Regulation

To ensure AI serves human flourishing rather than oppression, strong ethical and legal frameworks are essential. Three foundational principles stand out:

> 1. Transparency: Companies and governments should disclose, in accessible terms, how AI systems operate, how they are trained, and how they curate and distribute information.

> 2. Accountability: There must be clear responsibility for decisions influenced or made by AI, including avenues for redress when harm occurs.

> 3. Inclusivity: Regulations must address bias, protect vulnerable groups, and promote equitable access to the benefits of AI.

More than 70 countries and territories, along with the European Union, have adopted hundreds of policy initiatives addressing AI. The EU's Artificial Intelligence Act, which entered into force in August 2024, applies extraterritorially to any provider or deployer whose AI systems or outputs are used in the EU, regardless of where the company is based.[273] Among its prohibited practices are:

- Manipulative techniques that significantly impair informed decision-making or exploit specific vulnerabilities.

- Biometric categorization systems that infer sensitive attributes such as race, religion, or political beliefs.

- Remote biometric identification in publicly accessible spaces for law-enforcement purposes, except in narrowly defined circumstances.

- Systems that assess criminal risk based solely on profiling or personality traits.

The Act also classifies a range of applications—from critical infrastructure to employment and access to public services—as "high-risk," imposing strict requirements for risk management, data governance, documentation, and meaningful human oversight. Enforcement is coordinated by a European Artificial Intelligence Board, and serious breaches can attract fines of up to 7% of global annual turnover or €35 million, whichever is higher.

In contrast to hyper-surveillance models, such as that of the Chinese Communist Party, the EU framework is explicitly grounded in protecting fundamental rights and human dignity while enabling innovation.

In May 2019, the Organization for Economic Co-operation and Development (OECD) adopted high-level principles on trustworthy AI, emphasizing human-centred values, transparency, robustness, and accountability.[274] The Global Partnership on AI (GPAI) was launched by G7 leaders and partner countries in 2020 as a multistakeholder initiative to translate high-level AI principles into practice via working groups, research, and pilot projects. It was established with secretariat support from the OECD (and later UNESCO), but it is not formally "hosted at" the OECD; instead, the OECD provides administrative and analytic support to GPAI, which is legally and politically distinct. GPAI now includes several dozen participating and observer states, with experts from academia, industry, civil society, and government working together on topics such as responsible AI, data governance, and AI for the public good.[275]

Although global initiatives of this kind provide valuable forums for aligning ethical standards, the absence or withdrawal of key AI powers—whether through non-participation, reduced funding, or a shift toward unilateral dominance—risks fragmenting the emerging governance landscape.

The Global Digital Compact

In September 2024, UN member states adopted the Pact for the Future at the Summit of the Future in New York, which includes a Global Digital Compact (GDC).[276] The GDC aims to articulate shared principles for an open, free, and secure digital future, with several core objectives:

- Closing digital divides: Advancing universal, affordable connectivity and digital skills.

- Countering misinformation: Promoting access to trustworthy information and coordinated responses to misleading content.

- Governing AI responsibly: Strengthening international cooperation on AI regulation, safety, and accountability.

- Promoting human rights online: Reaffirming that existing human rights standards apply fully in the digital environment.

If implemented robustly, the Compact could help reduce regulatory gaps between states and set a baseline for ethical digital governance.

Shifting Regulatory Approaches in the United States

On 30 October 2023, President Joe Biden signed an Executive Order on the "Safe, Secure, and Trustworthy Development and Use of Artificial Intelligence," which directed federal agencies to assess and mitigate AI-related risks, including those to national security, civil rights, and labour markets.[277] Subsequent policy moves under President Donald Trump have placed greater rhetorical emphasis on AI innovation and US global competitiveness, and some commentators argue that these shifts downplay precautionary safeguards compared to the Biden administration's earlier approach.

While a lighter-touch approach may accelerate AI deployment in the short term, it also heightens risks associated with opaque and inadequately tested systems. It may also create friction for US-based companies that must simultaneously comply with more stringent regimes in the EU, UK, Canada, Japan, and other jurisdictions.

Controlling Neurotechnology: Preserving the Sanctity of the Mind

Neuroethics and Interdisciplinary Oversight

The rapid emergence of neurotechnology has generated profound ethical, legal, and social questions, giving rise to the field of neuroethics. It examines issues such as the privacy of neural data, the risks and benefits of brain–computer interfaces (BCIs), and the implications of cognitive enhancement.

Scholars such as Nita Farahany[278] and Rafael Yuste[279] argue that the brain is humanity's "last frontier" of autonomy, requiring special protection. Key neuroethical questions include:

- Who owns and controls the data generated from an individual's brain activity?

- To what extent should people be permitted to modify their own cognition, and under what conditions?

- What responsibilities do researchers, clinicians, and companies have to ensure that neurotechnologies are used ethically?

Interdisciplinary oversight is essential. Neuroethics brings together neuroscientists, clinicians, legal scholars, technologists, policymakers, and ethicists to craft guidelines that protect individual rights while enabling responsible scientific progress. Current recommendations include enhanced protocols for informed consent in BCI trials, rigorous data-security standards, careful monitoring of adverse effects, and transparent reporting of risks and benefits. Embedding neuroethics into each stage of research and deployment helps ensure that advances serve the common good and respect human dignity.

Cultural Considerations in Global Brain Initiatives

In 2018, a Global Neuroethics Summit published in *Neuron* a set of guiding questions for ethically responsible brain research, noting that many national initiatives had paid too little attention to cultural perspectives.[280] Authors from the United States, United Kingdom, Sweden, Japan, and Korea highlighted how differing views of mind, body, and self can shape ethical priorities. For example:

- In parts of China, mental illness is often framed as a family or societal issue rather than a strictly confidential doctor–patient matter, so preclinical diagnoses could stigmatize not only individuals but their relatives and communities.

- In some Buddhist contexts, brain banking may raise concerns that removing the brain after death disrupts the person's peaceful transition, complicating post-mortem donation.

- Large-scale pooling of brain-imaging data across borders increases the risk of re-identification, especially when combined with other datasets, challenging privacy even when data are nominally "anonymized."

Recognizing such cultural differences is crucial for building globally legitimate frameworks for neurotechnology.

Organoid Research

Brain organoids—miniaturized, lab-grown structures that model certain aspects of human brain development and activity—have transformed experimental neuroscience, enabling the study of cortical development and neurological disease in unprecedented detail. Yet as organoids become more complex, ethical questions intensify, particularly around transplanting human brain organoids into non-human animals.[281]

If chimeric animals were ever to display capacities that resemble human-like cognition, moral status and legal protections would need to be reconsidered. Debates now focus on how to define meaningful markers of consciousness or sentience in this context, and on drawing lines that prevent incremental drift toward ethically unacceptable experiments.

Cognitive Enhancement and Identity

Transhumanist advocates see human enhancement through AI, BCIs, and pharmacological interventions as a desirable next step in evolution. Critics warn that certain forms of integration could fundamentally erode what we understand as a human mind. Philosopher Susan Schneider, for example, has argued that fully merging with AI might amount to "suicide for the human mind," raising questions about continuity of identity if neural tissue is progressively replaced by artificial components.[282]

These debates are not merely speculative. As closed-loop BCIs and adaptive neurostimulators become more autonomous, they may begin to shape thoughts, moods, and behaviours in ways that blur the boundary between user and device, forcing societies to revisit notions of authenticity, consent, and selfhood.

Global and Societal Considerations

The Need for International Cooperation

Because AI and neurotechnologies easily cross borders, ethical "islands" are unlikely to be effective. States that adopt lax standards can become havens for risky experimentation or military applications. International bodies such as the International Atomic Energy Agency (IAEA) offer models for how powerful technologies can be monitored and constrained through inspections, reporting, and shared safety norms. A comparable framework for high-risk neurotechnologies and weaponizable AI may eventually be necessary.

Social Impact of Cognitive Enhancement

If cognitive-enhancing technologies are accessible only to wealthy individuals, corporations, or militaries, existing inequalities could deepen dramatically: enhanced elites might gain disproportionate control over resources and decision-making, while others feel pressured to accept risky interventions simply to remain competitive.[283] Such dynamics would not only threaten social cohesion but also undermine respect for unenhanced human capacities and diversity.

Safeguarding Autonomy and Privacy

The risk that hostile states, criminal networks, or rogue insiders might compromise neurotechnological systems underscores the need for rigorous oversight and security. Wireless neurodevices—such as deep brain stimulators or 'brain pacemakers'—are already known to be vulnerable to hacking; ethicists and clinicians have warned that malicious interference with these systems, sometimes dubbed 'brainjacking,' could in principle alter stimulation, disrupt therapy, or expose sensitive neural data.[284] In principle, attackers could alter stimulation parameters, interfere with therapies, or exfiltrate highly sensitive neural data.

To mitigate these risks, developers must implement strong encryption, authentication, and key-management protocols, moving away from weak proprietary systems toward open, independently evaluated security architectures. Regulatory agencies should require security audits and post-market monitoring as conditions of approval.

For neurotechnologies with clear, demonstrable benefits, deployment should be governed by strict ethical protocols grounded in several core principles:

- Informed consent: Patients and research participants must genuinely understand the risks, benefits, and long-term implications of neural interventions, including uncertainties.

- Data security and privacy: Neural data should be treated as among the most sensitive categories of personal information, protected against hacking, unauthorized surveillance, and secondary uses.

- Long-term oversight: Continuous monitoring is essential to ensure interventions remain safe, reversible where possible, and aligned with users' evolving interests and values.

For the most powerful applications, an international body of experts—analogous to the IAEA's inspection and reporting model—could help verify compliance, investigate abuses, and reduce the likelihood of a covert arms race in mind-affecting technologies.

Conclusion

The ethical challenges posed by mind-related technologies—from historical psychological, chemical, and surgical interventions to modern digital manipulation, AI systems, and neurotechnologies—are profound and far-reaching. As societies approach a future in which artificial intelligence and neural enhancement offer transformative benefits alongside serious risks, a rigorous ethical framework that prioritizes human dignity and autonomy becomes indispensable.

The cautionary lessons of past abuses show how easily unchecked power over the mind can be turned toward domination rather than care. Without robust oversight, transparent governance, and effective remedies, tools that could expand human capabilities may instead be deployed as instruments of control.

By committing to transparency, accountability, and international collaboration, humanity can strive to ensure that these technologies remain tools of liberation rather than mechanisms of subjugation. The words attributed to John Alston nearly two centuries ago still resonate: "If you don't control your mind, someone else will." The question for our time is sharper still: in an age of powerful AI and invasive neurotechnology, who will control the controllers—and under whose rules?

c. 27 BCE – Augustus establishes the principate and develops imperial propaganda to legitimize one-man rule and manage public opinion across the Roman world.

15th–17th centuries – Early print propaganda and religious polemic (Reformation pamphlets, state censorship, church indexes) show how mass media can steer belief and loyalty.

Late 18th–early 19th centuries – Revolutionary and Napoleonic wars use nationalist rhetoric, symbols, and mass conscription to fuse identity, obedience and military mobilization.

1914–1918 – World War I pioneers modern psychological warfare and total-war propaganda, alongside chemical weapons that terrorize troops and civilians.

Interwar period (1918–1939) – Soviet and Nazi regimes refine mass propaganda, radio, film, rallies, and secret-police methods as tools to control populations and crush dissent.

1932–1945 – Unit 731 and related Japanese units develop and field-test biological weapons on prisoners and Chinese civilians, merging human experimentation with weapons of mass destruction.

From 1938 – Electroconvulsive therapy (ECT) and psychosurgery are introduced; in some contexts they are used coercively, blurring treatment, punishment and behavioural control.

1945–early 1950s – Early Cold War psychological warfare units emerge; both blocs experiment with indoctrination, "brainwashing" narratives and exploitation of prisoners' vulnerabilities.

1953–1973 – CIA Project MKULTRA and related programs secretly test drugs, hypnosis, sensory deprivation and extreme ECT on often unwitting subjects in pursuit of interrogation and behaviour-modification techniques.

1960s–1970s – Modern neuromodulation begins, including deep-brain stimulation and early brain-implant experiments, raising fears and fantasies about direct technological mind control.

1990s–2000s – The internet and 24-hour news create a continuous information environment; states and non-state actors refine online propaganda, psy-ops and targeted influence campaigns.

2010s – Social media platforms, recommendation algorithms and social-bot networks enable personalized disinformation at scale, "astroturf" movements and election-focused influence operations.

2020s – Advancing AI, large-scale data capture and sophisticated neuromodulation converge, making fine-grained monitoring and manipulation of attention, emotion and behaviour increasingly feasible.

Referernces

The Anatomy of a Massacre

[1] Encyclopaedia Britannica. My Lai Massacre. 5 Feb
2026. https://www.britannica.com/event/My-Lai-
Massacrebritannica+1

Chapter 1

[2] UNIT 731 – Japan's Biological Warfare Project. Experiments. UNIT 731;
2023. : https://unit731.org/experiments/unit731+1

[3] Bowers JZ. Japanese biomedical experimentation during World War II.
In: *Military Medical Ethics, Volume 2*. Washington (DC): Borden
Institute; 2003. p. 513–
550. https://medcoeckapwstorprd01.blob.core.usgovcloudapi.ne
t/pfw-images/borden/ethicsvol2/Ethics-ch-16.pdf

[4] Amelia Hill, "Unit 731: Japan's Biological Force," *The Guardian*, August
31,

2002, https://www.theguardian.com/world/2002/aug/31/japan.am
hil.

[5] Daniel Barenblatt, *A Plague upon Humanity: The Secret Genocide of
Axis Japan's Germ Warfare Operation* (New York: HarperCollins,
2004), https://www.harpercollins.com/products/a-plague-upon-
humanity-daniel-barenblatt.harpercollins

[6] Prince Tsuneyoshi Takeda. In: *Wikipedia* [Internet]. Wikimedia
Foundation;
2024: https://en.wikipedia.org/wiki/Prince_Tsuneyoshi_Takeda

[7] Wallechinsky D. The biographies of all IOC members – Part XV: Prince
Tsuneyoshi Takeda. *Journal of Olympic History*.
2013;21(3): http://isoh.org/wp-content/uploads/JOH-
Archives/JOHv21n3t.pdf

[8] Krzzjn Editorial Committee. Unit 731. Harbin: Museum of Evidence of
War Crimes by Japanese Army Unit 731; 2009. p. 30–
33: https://www.krzzjn.com/book/135725/files/basic-
html/page30.htmlkrzzjn

[9] Krzzjn Editorial Committee. Unit 731. Harbin: Museum of Evidence of
War Crimes by Japanese Army Unit 731; 2009. p. 30–33.krzzjn
Nishiyama K, editor. https://m.krzzjn.com/show-2932-
135725.htmlkrzzjn

Chapter 2

[10] Sherif M. Experimental study of positive and negative intergroup
attitudes between experimentally produced groups: Robbers
Cave study. In: Sherif M, editor. *Intergroup Relations and
Leadership*. New York: Wiley; 1962. p. 103–134. See also an
accessible summary of the Robbers Cave
experiment: https://www.simplypsychology.org/robbers-
cave.html.simplypsychology

[11] Pollack Peacebuilding Systems. The foundations of intergroup conflict
research: The Robbers Cave experiment. Los Angeles (CA):
Pollack Peacebuilding;
2025. https://pollackpeacebuilding.com/blog/intergroup-conflict-
robbers-cave/

[12] Stanley Milgram, "Behavioral Study of Obedience," *Journal of Abnormal and Social Psychology* 67, no. 4 (1963): 371–378, https://www.columbia.edu/cu/psychology/terrace/w1001/readings/milgram.pdf.columbia+1

[13] Zimbardo PG. Stanford Prison Experiment. Philip G Zimbardo Official Site; 2024. : https://philipzimbardo.com/research/stanford-prison-experiment/

[14] Reicher SD, Haslam SA. Rethinking the psychology of tyranny: the BBC prison study. *British Journal of Social Psychology*. 2006;45(1):1–40. : http://www.bbcprisonstudy.org/pdfs/bjsp(2006)tyrannny.pdf

[15] Le Bon, Gustave. *The Crowd: A Study of the Popular Mind*. London: T. Fisher Unwin, 1896. Originally published in French in 1895. Accessed March 17, 2026. https://www.gutenberg.org/ebooks/445.gutenberg+1

[16] Le Bon G. The Crowd: A Study of the Popular Mind. London: T. Fisher Unwin; 1896. See esp. Ch. 1–2 for his account of the "collective mind" of crowds, anonymity, contagion, and suggestibility.

[17] Suler J. The online disinhibition effect. *CyberPsychology & Behavior*. 2004;7(3):321–326. : https://johnsuler.com/article_pdfs/online_dis_effect.pdf

[18] Bandura A. Moral disengagement in the perpetration of inhumanities. *Personality and Social Psychology Review*. 1999;3(3):193–209.

Chapter 3

[19] Propaganda in World War I. In: *Wikipedia* [Internet]. Wikimedia Foundation; 2026 [cited YEAR DATE]. : https://en.wikipedia.org/wiki/Propaganda_in_World_War_I

[20] Herf J. The Jewish Enemy: Nazi Propaganda during World War II and the Holocaust. Cambridge (MA): Harvard University Press; 2006. https://www.hup.harvard.edu/books/9780674027381hup.harvard

[21] United States Holocaust Memorial Museum. Holocaust Encyclopedia. Washington (DC): United States Holocaust Memorial Museum; c1998–[ongoing]. : https://encyclopedia.ushmm.org

[22] https://rarehistoricalphotos.com/nazi-rally-cathedral-light-c-1937/

[23] Landry AP, et al. "Dehumanization and mass violence: A study of mental state language in Nazi propaganda (1927–1945)." *PLOS ONE*. 2022;17(11):e0274957. : https://journals.plos.org/plosone/article?id=10.1371/journal.pone.0274957

[24] Artificial Intelligence and Political Deepfakes: Shaping Citizen Perceptions Through Misinformation
Mina Momeni https://orcid.org/0000-0002-9695-3708 mina.momeni@uwaterloo.ca, Volume 20, Issue 1

https://doi.org/10.1177/09732586241277335

Chapter 4

[25] Othering & Belonging Institute. The problem of othering: Towards inclusiveness and belonging. Berkeley (CA): University of California, Berkeley; : https://belonging.berkeley.edu/problem-othering-towards-inclusiveness-and-belongingbelonging.berkeley

[26] Blacker CP. *Eugenics: Galton and after*. Cambridge (MA): Harvard University Press; 1952. : The Online Books Page, University of Pennsylvania.
URL: https://onlinebooks.library.upenn.edu/webbin/book/lookupid?key=ha001355195onlinebooks.library.upenn

[27] Leonard T. "Retrospectives: Eugenics and economics in the Progressive Era." *Journal* of *Economic Perspectives*. 2005;19(4):207–224, https://www.hup.harvard.edu/books/9780674027381bookscouter +1

[28] Churchill WS. Letter to Prime Minister Herbert Asquith, Dec 1910. Quoted in: "Churchill and Eugenics." *Finest Hour* (Winston Churchill Centre); 2021.
: https://winstonchurchill.org/publications/finest-hour-extras/churchill-and-eugenics-1/

[29] Embryo Project Encyclopedia. Madison Grant (1865–1937). 19 June 2021. https://embryo.asu.edu/pages/madison-grant-1865-1937embryo.asu

30 National Human Genome Research Institute. Eugenics: Its origin and development (1883–present). 29 Nov 2021. https://www.genome.gov/about-genomics/educational-resources/timelines/eugenicsgenome

31 Parliamentary Education Office (Australia). Immigration Restriction Act 1901. 31 Oct 2024. https://peo.gov.au/understand-our-parliament/history-of-parliament/history-milestones/australian-parliament-history-timeline/events/immigration-restriction-act-1901 Parliamentary Education Office (Australia). Immigration Restriction Act 1901. 31 Oct 2024.

https://peo.gov.au/understand-our-parliament/history-of-parliament/history-milestones/australian-parliament-history-timeline/events/immigration-restriction-act-1901

32 Immigration History. Immigration Act of 1924 (Johnson-Reed Act). 31 Jan 2020. https://immigrationhistory.org/item/1924-immigration-act-johnson-reed-act/immigrationhistory Immigration History. Immigration Act of 1924 (Johnson-Reed Act). 31 Jan 2020. https://immigrationhistory.org/item/1924-immigration-act-johnson-reed-act/immigrationhistory

33 Indiana Historical Bureau. 1907 Indiana Eugenics Law. 3 Dec 2020. https://www.in.gov/history/state-historical-markers/find-a-marker/1907-indiana-eugenics-law/in Indiana Historical Bureau. 1907 Indiana Eugenics Law. 3 Dec 2020. https://www.in.gov/history/state-historical-markers/find-a-marker/1907-indiana-eugenics-law/in

34 Kaelber L. California Eugenics. University of Vermont; 14 Apr 2011. https://www.uvm.edu/~lkaelber/eugenics/CA/CA.htmluvm https://www.uvm.edu/~lkaelber/eugenics/CA/CA.htmluvm

35 Alberta Eugenics Board. In: *Wikipedia* [Internet]. 2024 update. https://en.wikipedia.org/wiki/Alberta_Eugenics_Boardwikipedia Alberta Eugenics Board. In: *Wikipedia* [Internet]. 2024 update. https://en.wikipedia.org/wiki/Alberta_Eugenics_Boardwikipedia

36 Binding K, Hoche A. *Die Freigabe der Vernichtung lebensunwerten Lebens: Ihr Maß und ihre Form*. Leipzig: Felix Meiner; 1920.

37 United States Holocaust Memorial Museum. The Biological State: Nazi racial hygiene, 1933–1939. Holocaust Encyclopedia. https://encyclopedia.ushmm.org/content/en/article/the-biological-state-nazi-racial-hygiene-1933-1939

38 United States Holocaust Memorial Museum. Euthanasia Program and Aktion T4. Holocaust Encyclopedia.

39 Broberg, Gunnar, and Nils Roll-Hansen, eds. *Eugenics and the Welfare State: Sterilization Policy in Denmark, Sweden, Norway, and Finland*. East Lansing: Michigan State University Press, 1996. https://openlibrary.org/books/OL9330627M/Eugenics_and_the_Welfare_State.openlibrary+1.

40 Weindling P. *Health, Race and German Politics between National Unification and Nazism, 1870–1945*. Cambridge: Cambridge University Press; 1989. https://www.cambridge.org/us/universitypress/subjects/history/history-medicine/health-race-and-german-politics-between-national-unification-and-nazism-1870-1945cambridge

41 United Nations DESA, Population Division. *International Migrant Stock 2024: Key facts and figures*. New York: United Nations; 2024. https://www.un.org/en/global-issues/migration (summary) and PDF: https://www.un.org/development/desa/pd/sites/www.un.org.development.desa.pd/files/undesa_pd_2025_intlmigstock_2024_key_facts_and_figures.pdf

42 American Sociological Association. Edward A. Ross. 17 Mar 2024. https://www.asanet.org/edward-a-ross/asanet American Sociological Association. Edward A. Ross. 17 Mar 2024. https://www.asanet.org/edward-a-ross/asanet

43 Theodore Roosevelt. "On American Motherhood." Speech to the National Congress of Mothers, Washington, 13 Mar 1905. Text at: http://digfir-published.macmillanusa.com/hewittlawson2e/hewittlawson2e_docs19_2.html

44 Oswald Mosley. In: *Wikipedia* [Internet]. https://en.wikipedia.org/wiki/Oswald_Mosleywikipedia Oswald Mosley. In: *Wikipedia* [Internet]. https://en.wikipedia.org/wiki/Oswald_Mosleywikipedia

45 "Rivers of Blood" speech. In: *Wikipedia* [Internet]. https://en.wikipedia.org/wiki/Rivers_of_Blood_speechwikipedia "Rivers of Blood" speech. In: *Wikipedia* [Internet]. https://en.wikipedia.org/wiki/Rivers_of_Blood_speechwikipedia

46 Raspail J. *Le Camp des Saints* [The Camp of the Saints]. Paris: Éditions Robert Laffont; 1973. Raspail J. *Le Camp des*

Saints [The Camp of the Saints]. Paris: Éditions Robert Laffont; 1973.

47 "The Camp of the Saints." In: *Wikipedia"The Camp of the Saints." In:*

    *Wikipedia*https://en.wikipedia.org/wiki/The_Camp_of_the_Saints wikipedia,
https://en.wikipedia.org/wiki/The_Camp_of_the_Saints.wikipedia

48 Great Replacement conspiracy theory. In: *Wikipedia* [Internet].
https://en.wikipedia.org/wiki/Great_Replacement_conspiracy_the ory Great Replacement conspiracy theory.
In: *Wikipedia* [Internet].
https://en.wikipedia.org/wiki/Great_Replacement_conspiracy_the ory

49 Unite the Right rally. In: *Wikipedia* [Internet].
https://en.wikipedia.org/wiki/Unite_the_Right_rallywikipedia

50 The Washington Post. Why Black people are afraid of "crazy" White people. 7 Jun 2022.
https://www.washingtonpost.com/opinions/2022/06/07/splc-poll-black-people-afraid-great-replacement/.

51 Southern Poverty Law Center. Racist "replacement" theory believed by half of Americans. 23 Jan 2025.
https://www.splcenter.org/resources/stories/poll-finds-support-great-replacement-hard-right-ideas/splcenter

Southern Poverty Law Center. Racist "replacement" theory believed by half of Americans. 23 Jan 2025.
https://www.splcenter.org/resources/stories/poll-finds-support-great-replacement-hard-right-ideas/splcenter

52 Bridge Initiative (Georgetown University). Global Islamophobia & the "Great Replacement" Conspiracy Theory. 31 Aug 2022.
https://bridge.georgetown.edu/research/global-islamophobia-the-great-replacement-conspiracy-theory/

53 NBC News (archived). The Buffalo supermarket shooting suspect posted an apparent manifesto repeatedly citing "Great Replacement" theory. 14 May 2022.
https://web.archive.org/web/20220515050536/https://www.nbcne ws.com/news/us-news/buffalo-supermarket-shooting-suspect-posted-apparent-manifesto-repeatedl-rcna28889web.archive+2

Chapter 5

54 Robert Jay Lifton, *Thought Reform and the Psychology of Totalism: A Study of "Brainwashing" in China* (New York: W. W. Norton, 1961; repr., Chapel Hill: University of North Carolina Press, 1989), https://uncpress.org/9780807842539/thought-reform-and-the-psychology-of-totalism/.uncpress

55 Thought reform in China.
In: Wikipedia, https://en.wikipedia.org/wiki/Thought_reform_in_China.

Chapter 6

56 John Birch Society. Encyclopaedia Britannica https://www.britannica.com/topic/John-Birch-Society.britannica

57 Fred C. Koch, *A Business Man Looks at Communism* (Wichita, KS: privately published, 1960). See also "Fred C. Koch," FBI file transcript, Internet Archive, https://archive.org/stream/FredC.Koch/fred-c-koch-fbi_djvu.txt.wikipedia+2

58 UnKoch My Campus. Funding report: Koch university funding update 2005–2019. http://www.unkochmycampus.org/funding-reportunkochmycampus

59 Center for Strategic Philanthropy and Civil Society (Duke University). John M. Olin Foundation (1953–2005). 24 Apr 2023. https://cspcs.sanford.duke.edu/cspcs-publication/john-m-olin-foundation-1953-2005/cspcs.sanford.duke

60 Intelexual Media. A short history of televangelism. 22 May 2025. https://www.intelexualmedia.com/post/a-short-history-of-televangelismintelexualmedia

61 Moral Majority," *Encyclopaedia Britannica*, https:// Paul Weyrich.
In: *Wikipedia* www.britannica.com/topic/Moral-Majority.britannica

62 Paul Weyrich," *Wikipedia*, https://en.wikipedia.org/wiki/Paul_Weyrich.

63 Robert J. Billings Is Dead at 68; Helped Form the Moral Majority." *The New York Times*. 1 June 1995.

https://www.nytimes.com/1995/06/01/us/robert-j-billings-is-dead-at-68-helped-form-the-moral-majority.html

64 Official Supreme Court opinion (District of Columbia v. Heller, 554 US 570 (2008)): https://supreme.justia.com/cases/federal/us/554/570/supreme.justia

65 Pew Research Center. "Key Facts About Americans and Guns." July 24, 2024. Accessed March 17, 2026. https://www.pewresearch.org/short-reads/2024/07/24/key-facts-about-americans-and-guns/.pewresearch+1

66 Federalist Society, "About Us," The Federalist Society https://fedsoc.org/about-us; "Federalist Society," *Encyclopaedia Britannica*, 2026,

https://www.britannica.com/topic/Federalist-Society; "Federalist Society," *Wikipedia*, last modified June 27, 2025, https://en.wikipedia.org/wiki/Federalist_Society.fedsoc+2

67 Sheldon Whitehouse, "The Third Federalist Society," speech, US Senate, February 19, 2024, https://www.whitehouse.senate.gov/news/speeches/the-third-federalist-society/.whitehouse.senate

68 Citizens for a Sound Economy," *Wikipedia*, 2026, https://en.wikipedia.org/wiki/Citizens_for_a_Sound_Econo my; "Tea Party movement," *Wikipedia* 2026, https://en.wikipedia.org/wiki/Tea_Party_movement.

69 Homeowners Affordability and Stability Plan Fact Sheet," US Department of the Treasury, February 18, 2009, https://home.treasury.gov/news/press-releases/20092181117388144; "Tea Party movement," *Wikipedia*, accessed March 7, 2026, https://en.wikipedia.org/wiki/Tea_Party_movement.treasury+1

70 Pew Research Center, "Trump's Staunch GOP Supporters Have Roots in the Tea Party," May 16, 2019, https://www.pewresearch.org/politics/2019/05/16/trumps-staunch-gop-supporters-have-roots-in-the-tea-party/ https://www.pewresearch.org/politics/2019/05/16/trumps-staunch-gop-supporters-have-roots-in-the-tea-party/; Joseph L. Wagner, "How the Tea Party Captured the GOP: Insurgent Factions in American Politics," *Journal of Economics and*

Politics 26, no. 1
(2018), https://collected.jcu.edu/jep/vol26/iss1/4.pewresearch+1

71 "Leonard Leo, Architect of Conservative Supreme Court Takes Wider
   Culture War Role," *National Catholic Reporter*, January 3,
   2024, https://www.ncronline.org/news/leonard-leo-architect-
   conservative-supreme-court-takes-wider-culture

72 "John Paul II New Evangelization Awardees," Catholic Information
   Center, https://cicdc.org/john-paul-ii-new-evangelization-
   awardees/; "Leonard Leo, Architect of Conservative Supreme
   Court, Takes on Wider Culture," *National Catholic Reporter*,
   January 3, 2024,
   CatholicCitizens.org, https://catholiccitizens.org/news/105574/leo
   nard-leo-architect-of-conservative-supreme-court-takes-on-
   wider-culture/.cicdc+1

73 Andy Kroll, Justin Elliott, and Joshua Kaplan, "Barre Seid Donated $1.6
   Billion to Conservative Marble Freedom Trust," *ProPublica*,
   August 21, 2022, https://www.propublica.org/article/dark-money-
   leonard-leo-barre-seid; Heidi Przybyla, "D.C. Attorney General Is
   Probing Leonard Leo's Network," *Politico*, August 21,
   2023, https://www.politico.com/news/2023/08/22/d-c-attorney-
   general-is-probing-leonard-leos-network-00112331; Tierney
   Sneed, "Senate Democrats Subpoena Leonard Leo in Supreme
   Court Ethics Probe," *CNN*, April 11,
   2024, https://www.cnn.com/2024/04/11/politics/leonard-leo-
   subpoena-supreme-court-ethics-probe

74 Dobbs v. Jackson Women's Health Organization," SCOTUSblog,
   accessed March 7,
   2026, https://www.scotusblog.com/cases/case-files/dobbs-v-
   jackson-womens-health-organization/; "Trump v. United States,"
   Legal Information Institute, Cornell Law School, accessed March
   7, 2026, https://www.law.cornell.edu/supremecourt/text/23-939;
   "Leonard Leo," *Encyclopaedia Britannica*, accessed March 7,
   2026, https://www.britannica.com/biography/Leonard-Leo

75 "Heritage Foundation," *Encyclopaedia Britannica*,
   2026, https://www.britannica.com/topic/Heritage-Foundation;
   "Mandate for Leadership," *Wikipedia*, 2026,

   https://en.wikipedia.org/wiki/Mandate_for_Leadership; Alex Noel,
   "New Conservatism: The Heritage Foundation," The American
   Leader, September 14,

2024, https://theamericanleader.org/timeline/new-conservatism-the-heritage-foundation/;

David Smith, "Trump's Administration Seems Chaotic, but He's Drawing Directly from Project 2025's Playbook," *The Conversation*, February 5, 2025, https://theconversation.com/trumps-administration-seems-chaotic-but-hes-drawing-directly-from-project-2025-playbook-248821.wikipedia+3

[76] "Project 2025 Publishes Comprehensive Policy Guide, 'Mandate for Leadership: The Conservative Promise,'" Heritage Foundation, April 20, 2023, https://www.heritage.org/press/project-2025-publishes-comprehensive-policy-guide-mandate-leadership-the-conservative-promise; "Project 2025," *Wikipedia*, https://en.wikipedia.org/wiki/Project_2025

[77] Center for Progressive Reform, "One Year of Project 2025: 53 Percent of Authoritarian Agenda Implemented," February 10, 2026, Project 2025 Executive Action Tracker, https://progressivereform.org/publications/one-year-of-project-2025-pr/;

[78] Gordon Gauchat, "Politicization of Science in the Public Sphere: A Study of Public Trust in the United States, 1974–2010," *American Sociological Review* 77, no. 2 (2012): 167–187, https://doi.org/10.1177/0003122412438225;

John Timmer, "Conservatives Lose Faith in Science over Last 40 Years," *Scientific American*, March 30, 2012, https://www.scientificamerican.com/article/conservatives-lose-faith-in-science-over-last-40-years/.

[79] "1. Trust in Government: 1958–2015," Pew Research Center, November 23, 2015,

https://www.pewresearch.org/politics/2015/11/23/1-trust-in-government-1958-2015/;

"Trust in Government," Pew Research Center, topic page, https://www.pewresearch.org/topic/politics-policy/trust-in-government/;

"Political Polarization in the American Public," Pew Research Center, June 12, 2014, https://www.pewresearch.org/politics/2014/06/12/political-polarization-in-the-american-public/.pewresearch+2

80 "The Polarization in Today's Congress Has Roots That Go Back Decades," Pew Research Center, March 10, 2022, https://www.pewresearch.org/short-reads/2022/03/10/the-polarization-in-todays-congress-has-roots-that-go-back-decades/;

"Partisan Antipathy: More Intense, More Personal," Pew Research Center, October 10, 2019, https://www.pewresearch.org/politics/2019/10/10/partisan-antipathy-more-intense-more-personal/.pewresearch+1

81 Daniel Cox, "A 'Scary' Survey Finding: 4 in 10 Republicans Say Political Violence May Be Necessary," NPR, February 11, 2021, https://www.npr.org/2021/02/11/966498544/a-scary-survey-finding-4-in-10-republicans-say-political-violence-may-be-necessa;

"America's Image Abroad Rebounds With Transition From Trump to Biden," Pew Research Center, June 10, 2021, https://www.pewresearch.org/global/2021/06/10/americas-image-abroad-rebounds-with-transition-from-trump-to-biden/.pewresearch+1

82 Pew Research Center, "U.S. Image Declines in Many Nations Amid Low Confidence in Trump," June 11, 2025, https://www.pewresearch.org/global/2025/06/11/us-image-declines-in-many-nations-amid-low-confidence-in-trump/; and Pew Research Center, "Confidence in Trump," June 11, 2025, https://www.pewresearch.org/global/2025/06/11/confidence-in-trump/.pewresearch+1

83 David E. Broockman and Joshua L. Kalla, "Love Fox? MSNBC? You May Be Locked in a 'Partisan Echo Chamber,'" University of California, Berkeley, April 20, 2023, https://news.berkeley.edu/2023/04/21/love-fox-msnbc-you-may-be-locked-in-a-partisan-echo-chamber-study-finds/;

Elizabeth A. Comer, "War of the Words: Political Talk Radio, the Fairness Doctrine, and Political Polarization in America" (honors thesis, University of Maine, 2012);

Ashani Amarasinghe and Paul A. Raschky, "Competing for Attention: The Effect of Talk Radio on Elections and Political Polarization in the US," 2022, arXiv:2206.13675; Chenyan Jia et al., "Reranking Partisan Animosity in Algorithmic Social Media Feeds Alters Affective

Polarization," *Science* (2025), https://doi.org/10.1126/science.ad u5584

Chapter 7

[84] "Case: United States v. Fred C. Trump, Donald Trump, and Trump Management, Inc.," Civil Rights Litigation Clearinghouse, accessed March 8, 2026.

"FinCEN Fines Trump Taj Mahal Casino Resort $10 Million for Significant and Long-Standing Anti-Money Laundering Violations," Financial Crimes Enforcement Network, March 5, 2015; Steve Reilly, "Exclusive: Trump's 3,500 Lawsuits Unprecedented for a Presidential Nominee," *USA Today*, June 2, 2016.

[85] Wayne Barrett, interview by Kelly McEvers, "'Village Voice' Reporter Recalls Roy Cohn's Early Influence on Trump," *NPR*, August 4, 2016, https://www.npr.org/2016/08/04/488722392/village-voice-reporter-recalls-roy-cohns-early-influence-on-trump.npr

[86] "Donald Trump Steps Up Claims the Election Is 'Rigged,'" NPR, October 16, 2016, https://www.npr.org/2016/10/17/498292048/donald-trump-steps-up-claims-the-election-is-rigged;

Nolan D. McCaskill, "Trump: I Could 'Shoot Somebody' and I Wouldn't Lose Voters," *Politico*, January 23, 2016, https://www.politico.com/story/2016/01/donald-trump-shooting-vote-218145.npr+1

[87] Donald J. Trump, campaign statement and speech, Mount Pleasant, SC, December 7, 2015, quoted in "Donald Trump Calls For 'Total and Complete Shutdown of Muslims Entering the United States,'" *Time*, December 7, 2015, https://time.com/4139476/donald-trump-shutdown-muslim-immigration/.pbs+4

[88] Federal Election Commission, *Official 2016 Presidential General Election Results*, FEC, 2017, https://www.fec.gov/resources/cms-content/documents/federalelections2016.pdf; see also "2016 United States Presidential Election," *270toWin*, https://www.270towin.com/2016-

election/ (for the 304–227 Electoral College tally and the popular-vote margin).

89 Donald J. Trump, tweets, November 27–28, 2016, quoted in Eric Bradner, "Election Fraud: Donald Trump Falsely Claims 'Millions' of Illegal Votes Cost Him Popular Vote," https://www.cnn.com/2016/11/27/politics/donald-trump-voter-fraud-popular-votecnn

90 Eric Trump, comments on *Hannity* (Fox News), June 6, 2017, discussed in "Eric Trump: Democrats in Washington Are 'Not Even People,'" *CNN*, June 7, 2017, https://edition.cnn.com/2017/06/07/politics/eric-trump-hannity-democrats-obstruction/index.html.edition.cnn+1

91 Glenn Kessler, Salvador Rizzo, and Meg Kelly, "Trump's False or Misleading Claims Total 30,573 over 4 Years," *Washington Post*, January 24, 2021, https://www.washingtonpost.com/politics/2021/01/24/trumps-false-or-misleading-claims-total-30573-over-four-years/.

92 Stephen D. Reicher and S. Alexander Haslam, "The Politics of Hope: Donald Trump as an Entrepreneur of Identity," *Scientific American*, November 19, 2016, https://www.scientificamerican.com/article/the-politics-of-hope-donald-trump-as-an-entrepreneur-of-identity/.news.uq.edu

93 "Louis Casiano, "Howard Stern Claims Trump 'Disgusted' by His Own Supporters: 'He Wouldn't Even Let Them in a F---ing Hotel,'" *Fox News*, May 12, 2020, https://www.foxnews.com/entertainment/howard-stern-claims-trump-disgusted-by-supporters.foxnews+1

94 Michael Cohen, *Disloyal: The True Story of the Former Personal Attorney to President Donald J. Trump* (New York: Skyhorse, 2020), and related television interviews in which Cohen said Trump viewed his supporters as "low class" and "pawns to be used for his benefit," e.g., MSNBC, December 10, 2022, https://www.youtube.com/watch?v=gSQRJ2np0O4.wikipedia

95 Alex Isenstadt, "Ex-Pence Aide Blasts Trump over Covid Response, Says She'll Vote for Biden," *Politico*, September 16, 2020, https://www.politico.com/news/2020/09/17/olivia-troye-ad-trump-coronavirus-417428.politico

96 Jonathan Chait, "Told by White House What to Say, Hannity Replies, 'Yes Sir,'" *New York Magazine Intelligencer*, April 24, 2022, points, https://nymag.com/intelligencer/2022/04/hannity-texts-meadows-white-house-told-what-to-say-fox-news.html.

97 "Australian Author Reveals Forgotten Donald Trump Tweet That Sums Up His Presidency," 7NEWS, June 4, 2024, https://7news.com.au/politics/donald-trump/australian-author-reveals-donald-trumps-forgotten-tweet-that-perfectly-sums-up-his-presidency-c-15009481.

98 Michael D. Shear and Karen Yourish, "How Trump Reshaped the Presidency in Over 11,000 Tweets," *New York Times*, November 2, 2019, and Karen Yourish et al., "Five Years, Thousands of Insults: Tracking Trump's Invective," *New York Times*, January 25, 2021; see also CBS News, "How Trump's 11,000+ Tweets Have Reshaped the Presidency," November 4, 2019, https://www.nytimes.com/interactive/2019/11/02/us/politics/trump-twitter-presidency.html; https://www.nytimes.com/2021/01/25/insider/Trump-twitter-insults-list.html; https://www.cbsnews.com/video/how-trumps-11000-tweets-have-reshaped-the-presidency/.nytimes

99 Stanley, Jason. *How Fascism Works: The Politics of Us and Them*. New York: Random House, 2018. Accessed March 17, 2026. https://www.penguinrandomhouse.com/books/586030/how-fascism-works-by-jason-stanley/.wikipedia+1

100 "Social Media Use by Donald Trump," *Wikipedia*, https://en.wikipedia.org/wiki/Social_media_use_by_Donald_Trump; Paul M. Barrett, Justin Hendrix, and Grant Sims, "How Tech Platforms Fuel US Political Polarization—and What Government Can Do About It," Brookings Institution, July 24, 2024, https://www.brookings.edu/articles/how-tech-platforms-fuel-u-s-political-polarization-and-what-government-can-do-about-it/.

Chapter 8

101 Pablo Barberá, "Social Media, Echo Chambers, and Political Polarization," working paper, accessed March 17, 2026, http://pablobarbera.com/static/echo-chambers.pdf.pablobarbera

102 Sarah Berman, *Don't Call It a Cult: The Shocking Story of Keith Raniere and the Women of NXIVM* (Toronto: Random House Canada, 2020), https://www.penguinrandomhouse.com/books/671288/dont-call-it-a-cult-by-sarah--berman/.penguinrandomhouse

103 Chantal Da Silva, "NXIVM: How Is Former Trump Adviser Roger Stone Connected to 'Cult' at Center of Sex-Trafficking Case?" *Newsweek*, May 1, 2018, https://www.newsweek.com/roger-stone-nxivm-sex-cult-allison-mack-keith-raniere-907558.

104 "NXIVM," *Wikipedia*, https://en.wikipedia.org/wiki/NXIVM.

105 "Jury Finds NXVIM Leader Keith Raniere Guilty of All Counts," US Attorney's Office, Eastern District of New York, June 19, 2019, https://www.justice.gov/usao-edny/pr/jury-finds-nxivm-leader-keith-raniere-guilty-all-counts.justice

106 "Make America Great Again," United States Patent and Trademark Office service-mark registration (serial no. 85783371), filed November 19, 2012, and registered July 14, 2015, for political use.

107 Jonathan Cohn, "How 'Make America Great Again' Explains Trump's Win," *The Atlantic*, November 9, 2016, https://www.theatlantic.com/politics/archive/2016/11/how-make-america-great-again-explains-trumps-win/507155/.ebsco

108 Jennifer Valentino-DeVries et al., "Several Members of Far-Right Groups, Including the Proud Boys, Joined Trump Supporters at the Capitol," *The New York Times*, January 11, 2021, https://www.nytimes.com/2021/01/11/us/capitol-attack-far-right-groups.html.

109 John Sides, "Figuring Out How Many 'MAGA Republicans' There Actually Are," *The Washington Post*, September 2, 2022, https://www.washingtonpost.com/politics/2022/09/02/trump-republicans-biden-maga/.washingtonpost

110 Garen J. Wintemute et al., "MAGA Republicans' Views of American Democracy and Society and Support for Political Violence in the United States: Findings from a Nationwide Population-Representative Survey," *PLOS ONE* 19, no. 1 (2024): e0295747, https://journals.plos.org/plosone/article?id=10.1371/journal.pone.0295747.journals.plos+1

111 Steven Hassan, *Combating Cult Mind Control: The #1 Best-selling Guide to Protection, Rescue, and Recovery from Destructive Cults* (Rochester, VT: Park Street Press, 2015). Accessed March 17, 2026. https://freedomofmind.com/combating-cult-mind-control/.freedomofmind+1

112 Federica Di Renzo, *Charisma and Cult in American Politics: A Political Sociology Case Study of MAGA* (Bachelor's thesis, LUISS University, Academic Year 2024–2025), https://tesi.luiss.it/44056/1/104482_DI%20RENZO_FEDERICA.pdf.tesi.luiss+1

113 Paul M. Barrett, Justin Hendrix, and Grant Sims, "How Tech Platforms Fuel US Political Polarization—and What Government Can Do About It," Brookings Institution, July 24, 2024, https://www.brookings.edu/articles/how-tech-platforms-fuel-u-s-political-polarization-and-what-government-can-do-about-it/.brookings

Chapter 9

114 "Stasi," *Wikipedia*, noting that the Ministry for State Security was founded on 8 February 1950 and grew into one of the most extensive and intrusive secret police organisations in the world, https://en.wikipedia.org/wiki/Stasi.

115 "East German Uprising of 1953," *Wikipedia*, https://en.wikipedia.org/wiki/East_German_uprising_of_1953.

116 "Stasi," *Wikipedia*, https://en.wikipedia.org/wiki/Stasi.

117Zersetzung," *Wikipedia*, https://en.wikipedia.org/wiki/Zersetzung.

- 118 "Zersetzung," *Wikipedia, The Free Encyclopedia* (describes Stasi Zersetzung as covert psychological repression using personalized surveillance-based interference, 'silent' repression, and social undermining rather than open terror), https://en.wikipedia.org/wiki/Zersetzung.wikipedia

119 "Federal Commissioner for the Records of the State Security Service of the former German Democratic Republic," Stasi Records Archive (BStU), https://www.stasi-unterlagen-archiv.de.

1. [120] "Informational Self-Determination," *Wikipedia*,
   https://en.wikipedia.org/wiki/Informational_self-determination.

Chapter 10

[121] Elisa Guerra-Doce, "Alcohol and Drugs in Prehistoric Europe:
   Archaeological Evidence of the Consumption of Psychoactive
   Substances," *Journal of Archaeological Method and Theory* 22,
   no. 3 (2015): 751–782; "Alcohol and Drugs: Not Just for Modern
   Man," Phys.org, May 11, 2014, https://phys.org/news/2014-05-
   alcohol-drugs-modern-anthropology-intoxication.html

[122] "Tassili Mushroom Figure," *Wikipedia*,
   https://en.wikipedia.org/wiki/Tassili_Mushroom_Figure.wikipedia

[123] Jerry B. Brown and Julie M. Brown, *The Psychedelic Gospels: The
   Secret History of Hallucinogens in Christianity* (Rochester, VT:
   Park Street Press, 2016). Accessed March 17,
   2026. https://www.innertraditions.com/books/the-psychedelic-
   gospels.

[124] Walter N. Pahnke, "Drugs and Mysticism: An Analysis of the
   Relationship between Psychedelic Drugs and the Mystical
   Consciousness" (PhD diss., Harvard University, 1963), available
   at the Multidisciplinary Association for Psychedelic Studies
   (MAPS)
   archive: https://maps.org/images/pdf/books/pahnke/.maps

[125] Roland R. Griffiths et al., "Mystical-type experiences occasioned by
   psilocybin mediate the attribution of personal meaning and
   spiritual significance 14 months later," *Journal of
   Psychopharmacology* 22, no. 6 (2008): 621–
   632, https://pmc.ncbi.nlm.nih.gov/articles/PMC3050654/.

[126] Kevin O. St. Arnaud and Donald Sharpe, "Entheogens and spiritual
   seeking: The quest for self-transcendence, psychological
   well-being, and psychospiritual growth," *Journal of Psychedelic
   Studies* 7, no. 1 (2023): 69–
   79, https://doi.org/10.1556/2054.2023.00206.

[127] Roland R. Griffiths et al., "Survey of Subjective 'God Encounter
   Experiences': Comparisons among Naturally Occurring
   Experiences and Those Occasioned by the Classic Psychedelics

Psilocybin, LSD, Ayahuasca, or DMT," *PLOS ONE* 14, no. 4 (2019): e0214377, https://doi.org/10.1371/journal.pone.0214377.

[128] Albert Hofmann, "History of the Basic Chemical Investigations on the Sacred Mushrooms of Mexico," https://beezone.com/hofmann_psilocybin.

[129] Jean Bassett Johnson, "The Elements of Mazatec Witchcraft," *Etnologiska Studier* 9 (1939): 85–119, https://www.samorini.it/doc1/alt_aut/ek/johnson-the-elements-of-mazatec-witchcraft.pdf.

[130] CIA document DOC_0000017374, available at https://documents.theblackvault.com/documents/mkultra/MKULTRA3/DOC_0000017374/DOC_0000017374.pdf; and discussion in "R. Gordon Wasson," https://en.wikipedia.org/wiki/R._Gordon_Wasson; and "Paul Blum," https://en.wikipedia.org/wiki/Paul_Blum.

[131] R. Gordon Wasson, "Seeking the Magic Mushroom," *Life* 42, no. 19 (May 13, 1957): 100–120,

https://www.wussu.com/shrooms/wasson_seeking_the_magic_mushroom_life_magazine_1957.pdf

[132] Albert Hofmann, "The active principles of the Mexican magic mushroom," *Journal of Pharmacy and Pharmacology* 11, no. 11 (1959): 581–583, and Albert Hofmann, *LSD: My Problem Child* (New York: McGraw-Hill, 1980)

[133] "Why Mushrooms, Part 4: The Debt We Owe to María Sabina," https://buffalocollective.co/why-mushrooms-part-4-the-debt-we-owe-to-maria-sabina/.buffalocollective+1

[134] Franz X. Vollenweider and John W. Smallridge, "Classic psychedelic drugs: Update on biological mechanisms," *Pharmacopsychiatry* 55, no. 3 (2022): 121–138, available at https://pmc.ncbi.nlm.nih.gov/articles/PMC9110100/.

[135] Laura Orsolini et al., "The 'Endless Trip' among the NPS Users: Psychopathology and Psychopharmacology in the Hallucinogen-Persisting Perception Disorder. A Systematic Review," *Frontiers in Psychiatry* 8 (2017): 240,

https://www.frontiersin.org/articles/10.3389/fpsyt.2017.00240/full.

[136] José M. Capriles et al., "Chemical evidence for the use of multiple psychotropic plants in a 1,000-year-old ritual bundle from South

America," *Proceedings of the National Academy of Sciences* 116, no. 23 (2019): 11207–11212 at: https://www.pnas.org/doi/10.1073/pnas.1902174116.

137 James Rucker et al., "Ayahuasca: A review of historical, pharmacological, and therapeutic aspects," *Psychiatry and Clinical Neurosciences Reports* 2, no. 3 (2023): e146, open-access at https://pmc.ncbi.nlm.nih.gov/articles/PMC11114307/.pmc.ncbi.nlm.nih+1

138 Rick Strassman, *DMT: The Spirit Molecule – A Doctor's Revolutionary Research into the Biology of Near-Death and Mystical Experiences* (Rochester, VT: Park Street Press, 2001). Overview and publication details at https://www.rickstrassman.com/publications/the-spirit-molecule/.simonandschuster+1

139 "History of LSD,"

https://en.wikipedia.org/wiki/History_of_LSD.wikipedia+3

Albert Hofmann, *LSD: My Problem Child* (New York: McGraw-Hill, 1980)

140 Hofmann, Albert. *LSD: My Problem Child.* 4th ed. New York: Multidisciplinary Association for Psychedelic Studies (MAPS), 2005. Accessed March 17, 2026. https://maps.org/books/lsd-my-problem-child-2nd-edition/

141 Martin A. Lee and Bruce Shlain, *Acid Dreams: The Complete Social History of LSD: The CIA, the Sixties, and Beyond* (New York: Grove Press, 1985). https://groveatlantic.com/book/acid-dreams/groveatlantic

142 Carl Bernstein, "The CIA and the Media," *Rolling Stone*, October 20, https://goodtimesweb.org/covert-operations/rs-bernstein-cia-media-oct-20-1977.html.cia+2

143 Jefferson Morley, "Luce Diaries Detail Her Bad LSD Trips — 'I Am Quite Gone,' Conservative Wrote," *Washington Post*, October 22, 1997, https://archive.seattletimes.com/archive/?date=19971022&slug=2567632.archive.seattletimes

144 Martin A. Lee and Bruce Shlain, *Acid Dreams: The Complete Social History of LSD: The CIA, the Sixties, and Beyond* (New York: Grove Press, 1985), summarized in the Los Angeles Review of Books: https://lareviewofbooks.org/article/the-rise-and-fall-of-lsd/.

[145] Abbie Hoffman, *Soon to Be a Major Motion Picture* (New York: G. P. Putnam's Sons, 1980), For a reprinted extract, see "Vol2pten2," http://www.splattgallery.com/Vol2pten2.html.splattgallery

[146] Public Law 90-639, "An Act to amend the Federal Food, Drug, and Cosmetic Act with respect to depressant and stimulant drugs, and for other purposes," 82 Stat. 1361 (October 24, 1968), https://www.govinfo.gov/content/pkg/STATUTE-82/pdf/STATUTE-82-Pg1361.pdf.

[147] Public Law 90-639 (Staggers–Dodd Amendments), enacted October 24, 1968

Chapter 11

[148] James S. Ketchum, *Chemical Warfare: Secrets Almost Forgotten – A Personal Story of Medical Testing of Army Volunteers with Incapacitating Chemical Agents During the Cold War (1955–1975)* (Santa Rosa, CA: ChemBook, 2006).archive+2 https://archive.org/details/chemicalwarfares0000ketcarchive

[149] https://www.intelligence.senate.gov/wp-content/uploads/2024/08/*sites*-default-files-hearings-95mkultra.pdf

[150] "Project MKUltra," *Wikipedia*, Web: https://en.*wikipedia*.org/wiki/*MKUltra*

[151] National Security Archive, "CIA Behavior Control Experiments Focus of New Scholarly Document Collection," December 23, 2024, https://nsarchive.gwu.edu/briefing-book/dnsa-intelligence/2024-12-23/cia-behavior-control-experiments-focus-new-scholarly

[152] Gordon Thomas, *Secrets & Lies: A History of CIA Mind Control & Germ Warfare* (JR Books, 2008). See description and editions via Goodreads: https://www.goodreads.com/book/show/2438730.Secrets_and_Lies.goodreads+2

[153] "Project MKUltra," *Wikipedia*, accessed March 17, 2026, https://en.wikipedia.org/wiki/MKUltra.

[154] Crowley, *Mid-Spectrum Incapacitant Programs*, citing Cabinet Defence Committee minutes, May 1963, approving increased R&D on lethal and incapacitating chemical agents and their dissemination.bradscholars.brad.achttps://bradscholars.brad.ac.uk/bitstream/handle/10454/709/ST_Report_No_6.pdf

[155] UK Parliament, written answer on Porton Down LSD trials, November 1994, cited in "The Bizarre True Story of When the UK Military Tested LSD on Royal Marines," *The Independent*, 23 May 2018.the-independent

[156] Defence Science and Technology Laboratory, "The Truth About Porton Down," 26 June 2016.wired-gov+1

Chapter 12

[157] Defence Science and Technology Laboratory, "The Truth About Porton Down," UK Ministry of Defence press release, 26 June 2016.gov+1

[158] Clare Dyer, "NHS Settles Claim of Patients Treated with LSD," *BMJ* 324 (2002): 383.pmc.ncbi.nlm.nih

[159] Statewatch, "Border Control for the 21st Century: 'Automated Passenger Processing'," *Statewatch* 16, no. 1 (2006), note on Porton Down LSD compensation; see also "Watch: When Royal Marines Were Given LSD," Forces News, 10 November 2021.statewatch+1, https://www.statewatch.org/publications/journal/vol-16-1-january-february-2006/

[160] Forces News. "Watch: When Royal Marines Were Given LSD." November 10, 2021. https://www.forcesnews.com/heritage/history/watch-when-royal-marines-were-given-lsd.forcesnews

[161] Ronald A. Sandison, "The Therapeutic Value of Lysergic Acid Diethylamide in Mental Illness," *Journal of Mental Science* 100, no. 419 (1954): 491–507; https://doi.org/10.1192/bjp.100.419.491

Ronald A. Sandison, "Further Studies in the Therapeutic Value of Lysergic Acid Diethylamide in Mental Illness," *Journal of Mental Science* 103, no. 431 (1957): 332–43.

162 Tom Peck, "The Bizarre True Story of When the UK Military Tested LSD on Royal Marines," *Independent*, May 23, 2018, https://www.independent.co.uk/news/uk/home-news/lsd-video-porton-down-chemical-weapons-experiments-trials-uk-military-army-marines-a8360216.html.

163 William Walters Sargant and Eliot Slater, *An Introduction to Physical Methods of Treatment in Psychiatry* (Edinburgh: E. & S. Livingstone, 1944).worldcat+2

164 William Sargant, *Battle for the Mind: A Physiology of Conversion and Brain-Washing* (London: Heinemann, 1957), https://malorbooks.com/titles/battle-for-the-mind-william-sargant/; William Sargant, *The Unquiet Mind: The Autobiography of a Physician in Psychological Medicine* (London: Heinemann, 1967); William Sargant, *The Mind Possessed: A Physiology of Possession, Mysticism, and Faith Healing* (London: Heinemann, 1973).

Chapter 13

165 David S. Kushner, John W. Verano, and Anne R. Titelbaum, "Trepanation Procedures/Outcomes: Comparison of Prehistoric Peru with Other Ancient, Medieval, and American Civil War Cranial Surgery," *World Neurosurgery* 114 (2018): 245–254. https://pubmed.ncbi.nlm.nih.gov/29604358/pubmed.ncbi.nlm.nih

166 António Egas Moniz, "Prefrontal Leucotomy in the Treatment of Mental Disorders," *American Journal of Psychiatry* 93 (1936): 137–156.

167 Ernest A. Spiegel and Henry T. Wycis, "Stereoencephalotomy: Part I. Methodology," *Journal of the International College of Surgeons* 11 (1948): 537–550; and Ernest A. Spiegel and Henry T. Wycis, *Stereoencephalotomy, Part II: Clinical Aspects* (New York: Grune & Stratton, 1952).

168 Walter Freeman and James W. Watts, *Psychosurgery: Intelligence, Emotion and Social Behavior Following Prefrontal Lobotomy for Mental Disorders* (Springfield, IL: Charles C. Thomas, 1942).

169 Walter Freeman and James W. Watts, *Psychosurgery: Intelligence, Emotion and Social Behavior Following Prefrontal Lobotomy for Mental Disorders* (Springfield, IL: Charles C. Thomas, 1942).

170 Timothy C. Marzullo, "The Missing Manuscript of Dr. Jose Delgado's Radio Controlled Bulls," *Journal of Undergraduate Neuroscience Education* 15, no. 2 (June 15, 2017): R29–R35. https://pmc.ncbi.nlm.nih.gov/articles/PMC5480854/pmc.ncbi.nlm.nih

171 José Manuel Rodríguez Delgado, Wikipedia; https://en.wikipedia.org/wiki/Jos%C3%A9_Manuel_Rodr%C3%ADguez_Delgado.wikipedia

- 172 Nicholas D. Lorusso, Uma R. Mohan, and Joshua Jacobs, "Jose Delgado: A Controversial Trailblazer in Neuromodulation," *Artificial Organs* (2022), accessed March 17, 2026, https://onlinelibrary.wiley.com/doi/10.1111/aor.14200.onlinelibrary.wiley+1

173 Jérôme Munuera, Social Hierarchy Representation in the Primate Amygdala Reflects the Emotional Ambiguity of Our Social Interactions, J Exp Neurosci . 2018 Jun 17;12:1179069518782459. doi: 10.1177/1179069518782459

174 Brain Researcher José Delgado Asks— 'What Kind of Humans Would We Like to Construct?' - By Maggie Scarf Nov. 15, 1970; https://www.nytimes.com/1970/11/15/archives/brain-researcher-jose-delgado-asks-what-kind-of-humans-would-we.html

175 "Robert Galbraith Heath." *Wikipedia, The Free Encyclopedia.* Last modified 2024 (check page for current date). Available at: https://en.wikipedia.org/wiki/Robert_Galbraith_Heathwikipedia

176 Lisman, John E., et al. "Neuroimaging of frontal–limbic dysfunction in schizophrenia and epilepsy-related psychosis: Toward a convergent neurobiology." *Epilepsy & Behavior* 23(2), 2012: 113–122: https://pmc.ncbi.nlm.nih.gov/articles/PMC3339259/pmc.ncbi.nlm.nih

177 Robert Colvile, "The 'gay cure' experiments that were written out of scientific history," *Mosaic*, July 4, 2016, *http://mosaicscience.com/story/gay-cure-experiments*,

- 178 Elliot S. Valenstein, *Brain Control: A Critical Examination of Brain Stimulation and Psychosurgery* (New York: Wiley-Interscience, 1973), accessed March 17,

2026, https://archive.org/details/braincontrol0000vale.catalog.nlm.nih
+2

[179] CIA Reading Room – "PROJECT MKULTRA" main file:
https://www.cia.gov/readingroom/document/06760269web.archive

[180] Violence and the brain : Mark, Vernon H., 1925– ; Ervin, Frank R."
Internet Archive catalog page, confirming co-authorship and 1970
publication: https://archive.org/details/violencebrain0000mark

[181] Beran, Roy G. "Episodic Dyscontrol, Rage, and Violence."
In *Borderland of Epilepsy Revisited*. Oxford: Oxford University
Press, 2012.

https://doi.org/10.1093/med/9780199796793.003.0020.

[182] Shaw J, Pyreddy S, et al. Current Neuroethical Perspectives on Deep
Brain Stimulation and Neuromodulation for Neuropsychiatric
Disorders: A Scoping Review of the Past 10
years. *Diseases*. 2025;13(8):262.
doi:10.3390/diseases13080262. Available
at: https://pubmed.ncbi.nlm.nih.gov/40863235/pubmed.ncbi.nlm.
nih+1

Chapter 15

[183] Mayberg HS, Lozano AM, Voon V, et al. "Deep Brain Stimulation for
Treatment-Resistant Depression." *Neuron*. 2005 Mar
3;45(5):651–660.
https://pubmed.ncbi.nlm.nih.gov/15748841/

[184] Holtzheimer PE, Kelley ME, Gross RE, et al. "Subcallosal cingulate
deep brain stimulation for treatment-resistant unipolar and
bipolar depression." *Arch Gen Psychiatry*. 2012;69(2):150–158.
Open summary via
PubMed/PMC: https://pubmed.ncbi.nlm.nih.gov/22213770/pubm
ed.ncbi.nlm.nih+1

[185] FDA HDE approval record: Humanitarian Device Exemption H050003
– Medtronic Reclaim DBS Therapy for OCD. US Food and Drug
Administration. Approval order and indication statement (anterior
limb of internal capsule, chronic severe treatment-resistant OCD;

approval date 19 February 2009).
https://www.accessdata.fda.gov/scripts/cdrh/cfdocs/cfhde/hde.cfm?id=375533
Summary of safety and probable benefit (PDF):
https://www.accessdata.fda.gov/cdrh_docs/pdf5/H050003b.pdf

[186] Closed-Loop Deep Brain Stimulation for Refractory Chronic Pain, Prasad Shirvalkar et al, Front Comput Neurosci, . 2018 Mar 26;12:18. doi: 10.3389/fncom.2018.00018

[187] Voon V, Krack P, Lang AE, et al. "Psychiatric Complications of Deep Brain Stimulation for Parkinson's Disease." *Journal of Clinical Psychiatry.* 2004;65(6):XXX–XXX.

Chapter 16

[188] Turing, A. M. "Computing Machinery and Intelligence." *Mind*, New Series, 59, no. 236 (October 1950): 433–460. Available at: https://academic.oup.com/mind/article/LIX/236/433/986238

[189] History of Artificial Intelligence." *Wikipedia, The Free Encyclopedia.* https://en.wikipedia.org/wiki/History_of_artificial_intelligence

[190] Arute, F. et al. "Quantum supremacy using a programmable superconducting processor." *Nature* 574, 505–510 (2019). "Sycamore (processor)." *Wikipedia, The Free Encyclopedia.* *https://pubmed.ncbi.nlm.nih.gov/31645734/pubmed.ncbi.nlm.nih*

[191] "Neuromorphic computing." *Wikipedia, The Free Encyclopedia.*

[192] "Loihi – Intel neuromorphic research chip." Intel Labs technical overview (2018 onward).

[193] Kagan, B. J., Kitchen, A. C., et al. "In vitro neurons learn and exhibit sentience when embodied in a simulated game-world." Published 2022.

[194] Strong V, Hayashi Y, et al. "Ionic hydrogels can play Pong by 'remembering' previous patterns of electrical stimulation." *Cell Reports Physical Science.* 2024;5(8):102151.

[195] "AI bot, ChaosGPT tweet plans to 'destroy humanity' after being tasked." *New York Post*, 11 April 2023.

https://nypost.com/2023/04/11/ai-bot-chaosgpt-tweet-plans-to-destroy-humanity-after-being-tasked/nypost

196 Blaise Agüera y Arcas, "Artificial neural networks are making strides towards consciousness," *The Economist*, June 11, 2022 (By Invitation); https://www.economist.com/by-invitation/2022/09/02/artificial-neural-networks-are-making-strides-towards-consciousness-according-to-blaise-aguera-y-arcas

Chapter 17

197 Sorin, A., Brin, S., Klang, E., et al. "Large Language Models and Empathy: Systematic Review." *Journal of Medical Internet Research* 2024;26:e52597.jmir+3; https://www.jmir.org/2024/1/e52597/jmir

198 Good, I. J. "Speculations Concerning the First Ultraintelligent Machine." *Advances in Computers* 6 (1965): 31–88. https://www.sciencedirect.com/science/chapter/bookseries/pii/S0065245808604180sciencedirect

199 Vernor Vinge, "The Coming Technological Singularity: How to Survive in the Post-Human Era," *Whole Earth Review*, no. 77, Winter 1993. https://edoras.sdsu.edu/~vinge/misc/singularity.htmledoras.sdsu

200 Kurzweil, Ray. *The Singularity Is Near: When Humans Transcend Biology.* New York: Viking, 2005. https://books.google.com/books/about/The_Singularity_is_Near.html?id=88U6hdUi6D0Cbooks.google

201 Kurzweil, Ray. *The Singularity Is Nearer: When We Merge with AI.* New York: Viking, 2024. https://www.thekurzweillibrary.com/book-the-singularity-is-nearerthekurzweillibrary

202 Stephen Hawking, Stuart Russell, Max Tegmark, and Frank Wilczek, "Transcending Complacency on Superintelligent Machines," *The Huffington Post*, April 2014 (originally published as "Are We Taking AI Seriously Enough?"). https://www.independent.co.uk/news/science/stephen-hawking-transcendence-looks-at-the-implications-of-artificial-intelligence-

but-are-we-taking-ai-seriously-enough-9313474.htmlindependent+1

203 Harari, Yuval Noah. "How to Survive the 21st Century." Speech at the World Economic Forum Annual Meeting, Davos, 21–24 January 2020. Published by the World Economic Forum, 23 January 2020.weforum+2YouTube

- 204 European Parliamentary Research Service, "The Ethics of Artificial Intelligence: Issues and Initiatives," Study for the Panel for the Future of Science and Technology (STOA), European Parliament, March 2020, document code EPRS_STU(2020)634452. https://www.europarl.europa.eu/thinktank/en/document/EPRS_STU(2020)634452[1]

205 Deutsche Welle, "China's Social Credit System Experiment" (2018). https://www.dw.com/en/china-experiments-with-sweeping-social-credit-system/a-42030727

206 Human Rights Watch, *China's Algorithms of Repression: Reverse Engineering a Xinjiang Police Mass Surveillance App* (2019). https://www.hrw.org/report/2019/05/01/chinas-algorithms-repression/reverse-engineering-xinjiang-police-mass

Chapter 18

207 Amnesty International, *"Like We Were Enemies in a War": China's Mass Internment, Torture and Persecution of Muslims in Xinjiang* (2021). https://www.amnesty.org/en/documents/asa17/4137/2021/en/

208 Metz, Cade. "Meet GPT-3. It Has Learned to Code (and Blog and Argue)." *The New York Times*, November 24, 2020.nytimes; https://www.nytimes.com/2020/11/24/science/artificial-intelligence-ai-gpt3.htmlnytimes Roose, Kevin. "How Do You Know a Human Wrote This?" *The New York Times*, July 29, 2020.nytimes; https://www.nytimes.com/2020/07/29/opinion/gpt-3-ai-automation.html

209 Hern, Alex. "Forget Deepfakes – We Should Be Very Worried about AI-Generated Text." *The Guardian / The Telegraph* style opinion, August

2020, https://www.theguardian.com/technology/2020/aug/13/forg
et-deepfakes-we-should-be-very-worried-about-ai-generated-text

[210] Abid, Abubakar, Maheen Farooqi, and James Zou. "Persistent Anti-Muslim Bias in Large Language Models." *Proceedings of the 2021 ACM Conference on Fairness, Accountability, and Transparency (FAccT)*, 2021.
https://dl.acm.org/doi/10.1145/3461702.3462624

[211] Bender, Emily M., Timnit Gebru, Angelina McMillan-Major, and Shmargaret Shmitchell. "On the Dangers of Stochastic Parrots: Can Language Models Be Too Big?" Proceedings of the 2021 ACM Conference on Fairness, Accountability, and Transparency (FAccT '21). https://s10251.pcdn.co/pdf/2021-bender-parrots.pdfs10251.pcdn

Chapter 19

[212] Obama, Barack. "Remarks by the President on the BRAIN Initiative and American Innovation." The White House, 2 April 2013.
https://obamawhitehouse.archives.gov/the-press-office/2013/04/02/remarks-president-brain-initiative-and-american-innovation

[213] European Commission, "The Human Brain Project – Project Factsheet (Abstract & Partners)," 2014. https://digital-strategy.ec.europa.eu/en/library/human-brain-project-project-factsheet-abstract-partners

[214] Okano, Hideyuki. "Brain/MINDS: Brain-Mapping Project in Japan." *Neuron* 87, no. 1 (2015): 121–134. pmc.ncbi.nlm.nih.
https://pmc.ncbi.nlm.nih.gov/articles/PMC4387516/

[215] Kringelbach, M. L., et al. "Deep Brain Stimulation for Neurological and Psychiatric Disorders: Progress, Challenges and Opportunities." *Nature Reviews Neuroscience* 16, no. 8 (2015): 487–497. https://pubmed.ncbi.nlm.nih.gov/26077761/

[216] Musk, Elon, DJ Seo, Paul Merolla, et al. "An Integrated Brain-Machine Interface Platform with Thousands of Channels." Preprint, 2019. (Often cited via bioRxiv). https://www.biorxiv.org/content/10.1101/703801v4.full.pdf

217 Oxley, Thomas J., et al. "Assessment of Safety of a Fully Implanted Endovascular Brain-Computer Interface for Severe Paralysis in 4 Patients: The Stentrode With Thought-Controlled Digital Switch (SWITCH) Study." *JAMA Neurology* 80, no. 5 (2023): 533–541. PubMed: https://pubmed.ncbi.nlm.nih.gov/36622685/pubmed.nc bi.nlm.nih+1

218 Magnetogenetic systems, for example, use specially engineered cells that respond to magnetic fields, allowing external fields to activate or inhibit specific circuits in animal models. https://www.frontiersin.org/articles/10.3389/fncel.2024.1459120/f ullpmc.ncbi.nlm.nih

219 Deisseroth, Karl. "Optogenetics: Using Light to Control the Brain." *Scientific American* / reprinted in *Brain Research Bulletin* and NIH re-hosted article. : https://pmc.ncbi.nlm.nih.gov/articles/PMC3574762/pmc. ncbi.nlm.nih

220 "Neuralink: Musk's firm says first brain-chip patient plays online chess." *BBC News*, 20 March 2024. Web: https://www.bbc.com/news/business-68622781bbc

221 Fernández, Eduardo, et al. "Blind Volunteers Test Visual Neuroprosthesis That Communicates with the Brain's Visual Cortex." *Science Advances*-reported study, summarized by MedicalXpress, 2025. Web summary:

https://medicalxpress.com/news/2025-11-volunteers-visual-neuroprosthesis-communicates-brain.html

222 "The types of devices that Troyk and Neuralink are building bypass the eye and optic nerve completely, sending information straight to the brain."wired Web: https://www.wired.com/story/the-next-frontier-for-brain-implants-is-artificial-vision-neuralink-elon-musk/

223 Ezzyat, Youssef, et al. "Closed-Loop Stimulation of Temporal Cortex Rescues Functional Networks and Improves Memory." *Nature Communications* 9, no. 365 (2018).pmc.ncbi.nlm.nih+2 Web: https://pubmed.ncbi.nlm.nih.gov/29410414/ (abstract)pubm ed.ncbi.nlm.nih Full text: https://pmc.ncbi.nlm.nih.gov/articles/PMC5802791/pmc.ncbi .nlm.nih

Chapter 20

[224] Deppe, Ulrike, and Holger Schaal. "Cognitive Warfare: A Conceptual Analysis of the NATO ACT Cognitive Warfare Concept." *Frontiers in Big Data* 7 (2024): 1452129.
Web: https://www.frontiersin.org/articles/10.3389/fdata.2024.1452129/full

[225] Scales, Robert H. *Scales on War: The Future of America's Military at Risk.* Annapolis, MD: Naval Institute Press, 2016.usni+1.
Publisher page with details: US Naval Institute Press book page: https://www.usni.org/press/books/scales-warusni

[226] NATO Allied Command Transformation – "Cognitive Warfare" (official explainer), Web: https://www.act.nato.int/activities/cognitive-warfare/act.nato+1

[227] Pariser, Eli. *The Filter Bubble: What the Internet Is Hiding from You.* New York: Penguin Press, 2011.
(General info/overview:

https://books.google.com/books/about/The_Filter_Bubble.html?id=Qn2ZnjzCE3gC)

[228] Badawy, A., Ferrara, E., & Lerman, K. "Disinformation Warfare: Understanding State-Sponsored Trolls on Twitter and Their Influence on the Web."
Web (PDF): https://seclab.bu.edu/papers/trolls-cybersafety2019.pdf

[229] AFP Fact Check – Biden draft deepfake
Web: https://factcheck.afp.com/doc.afp.com.33YP34Mfactcheck.afp

Euronews / NPR – Zelenskyy surrender deepfake
Web (Euronews): https://www.euronews.com/my-europe/2022/03/16/deepfake-zelenskyy-surrender-video-is-the-first-intentionally-used-in-ukraine-wareuronews
Web
(NPR): https://www.npr.org/2022/03/16/1087062648/deepfake-video-zelenskyy-experts-war-manipulation-ukraine-russia

[230] NewsGuard special report (primary reference)
https://www.newsguardtech.com/special-reports/moscow-based-

global-news-network-infected-western-artificial-intelligence-russian-propaganda/newsguardtech+1

[231] Web: IMD, "Psychographics: the behavioural analysis that helped Cambridge Analytica" – https://www.imd.org/research-knowledge/technology-management/articles/psychographics-the-behavioural-analysis-that-helped-cambridge-analytica/imd

[232] Oxford Internet Institute / DemTech (Woolley & Howard) Web: https://navigator.oii.ox.ac.uk/what-is-comprop/navigator.oii.ox

European Parliament briefing on computational propaganda techniques https://epthinktank.eu/2018/10/22/computational-propaganda-techniques/epthinktank

[233] Howard, Philip N., et al. "Computational Propaganda Worldwide: Executive Summary." Oxford Internet Institute, 2017. Web (PDF): https://demtech.oii.ox.ac.uk/wp-content/uploads/sites/12/2017/06/Casestudies-ExecutiveSummary.pdfdemtech.oii.ox

[234] Web: "Disinformation and Civil Defence: How did Taiwan's civil society counter foreign information manipulation in the 2024 election?" – https://taiwaninsight.org/2024/02/05/disinformation-and-civil-defence-how-did-taiwans-civil-society-counter-foreign-information-manipulation-in-the-2024-election/

Chapter 21

Web: "Taiwan's Fact-Checking Ecosystem Makes Impression on International Visitors" – https://en.tfc-taiwan.org.tw/en_tfc_246/tfc-taiwan

Web (Quartz): "Taiwan is using humor as a tool against coronavirus hoaxes" – https://qz.com/1863931/taiwan-is-using-humor-to-quash-coronavirus-fake-newsqz

[235] Compton, J., van der Linden, S., Cook, J., & Basol, M. (2021). "Inoculation theory in the post-truth era: Extant findings and new frontiers for contested science, misinformation, and conspiracy theories." *Social and Personality Psychology Compass.*

Web: https://compass.onlinelibrary.wiley.com/doi/full/10.1111/spc
3.12602

[236] NATO Allied Command Transformation – "Beyond Military Information
Support Operations"
Web: https://www.act.nato.int/article/cognitive-warfare-beyond-
military-information-support-operations/

[237] AI-Driven Information Warfare: Disinformation and Psychological
Manipulation, Technology, Aryamehr Fattahi | 8 December 2025,
https://bisi.org.uk/reports/ai-driven-information-warfare-
disinformation-and-psychological-manipulationbisi.org

[238] ISW – "A Primer on Russian Cognitive Warfare" (2025)
Web: https://understandingwar.org/backgrounder/primer-russian-
cognitive-warfareunderstandingwar+1

Chapter 22

[239] Paziuk, Andrii, Dmytro Lande, Elina Shnurko-Tabakova, and Phillip
Kingston. "Decoding Manipulative Narratives in Cognitive
Warfare: A Case Study of the Russia-Ukraine Conflict." *Frontiers
in Artificial Intelligence* 8 (2025): 1566022.frontiersin+1
Web: https://www.frontiersin.org/articles/10.3389/frai.2025.15660
22/fullfrontiersin+1

[240] Dhiman, Ankit. "AI-Driven Disinformation Campaigns on Twitter (X) in
the Russia-Ukraine War." *Small Wars Journal* (2025).
Web: https://irregularwarfare.org/uncategorized/ai-driven-
disinformation-campaigns-on-twitter-x-in-the-russia-ukraine-war/

[241] Briggs, Chad M., and Anita Tusor. "Surveying New Battlegrounds:
Ukraine and the Future of Cognitive Warfare." *Journal of
Strategic Security* 18, no. 4 (2025).
Web: https://digitalcommons.usf.edu/jss/vol18/iss4/12/digitalcom
mons.usf+1

[242] Kluger, Jeffrey. "Ukraine's 'Secret Weapon' Against Russia Is
Clearview AI." *Time* (2023).
Web: https://time.com/6334176/ukraine-clearview-ai-russia/time

Newman, Lily Hay. "Ukraine Is Scanning Faces of Dead Russians, Then
Contacting Their Mothers." *The Washington Post* (2022).

Web: https://www.washingtonpost.com/technology/2022/04/15/ukraine-facial-recognition-warfare/

243 Bowe, Alexander. "Cognitive Domain Operations: The PLA's New Holistic Concept for Influence Operations." *Jamestown Foundation, China Brief* (2025). web: https://jamestown.org/cognitive-domain-operations-the-plas-new-holistic-concept-for-influence-operations/jamestown+1

244 Biddle, Sam. "US Special Forces May Use Deepfake Technology for Psy-Ops, Report Says." *The Independent* (2023). Web: https://www.independent.co.uk/news/world/americas/us-special-forces-deepfake-misinformation-b2296080.html

245 Keinan, Tom. "Israel's Digital Battlefield: Battling AI Disinformation." *The Jerusalem Post* (2025). Web: https://www.jpost.com/opinion/article-876478jpost

246 Frenkel, Sheera, Mark Mazzetti, and Ronen Bergman. "Israel Secretly Targets US Lawmakers with Influence Campaign on Gaza War." *The New York Times* (2024). Web: https://www.nytimes.com/2024/06/05/technology/israel-campaign-gaza-social-media.html

247 Maher, Shiraz. "Digital Frontlines: What the 12-Day War Revealed About the Evolution of Iran's Cyber Strategy." *Middle East Institute* (2025). Web: https://mei.edu/publication/digital-frontlines-what-12-day-war-revealed-about-evolution-irans-cyber-strategy/

248 Iran International. "Iran-Linked Influence Campaign Targeted Western Debate on Uprising." *Iran International* (2026). Web: https://www.iranintl.com/en/202601261504

249 Rauschenbach, Kyra. "Reality Defender Joins NATO's Cognitive Warfare Experimentation." *Reality Defender* (2026). Web: https://www.realitydefender.com/insights/reality-defender-joins-nato--cognitive-warfare-experimentation

250 DiResta, Renée, et al. "Researchers Caught a Pro-US Campaign Spreading Propaganda on Social Media." *The New York Times* (2022). Web: https://www.nytimes.com/2022/08/25/technology/facebook-twitter-influence-campaign.htmltheverge+1

251 DARPA. "N3: Next-Generation Nonsurgical Neurotechnology." *Defense Advanced Research Projects*

Agency (program description, updated 2020s).
Web: https://www.darpa.mil/research/programs/next-generation-nonsurgical-neurotechnology

252 Reilly, Katie. "The US Military Creates Tech to Control Drones With Thoughts." *Big Think* (2018). Web: https://bigthink.com/the-present/the-u-s-military-is-developing-mind-controlled-drones/

253 Gordon, Emma R., and Anil Seth. "The Ethics of AI-Assisted Warfighter Enhancement Research and Experimentation: Historical Perspectives and Ethical Challenges." *Frontiers in Big Data* 5 (2022): 978734.
Web: https://pmc.ncbi.nlm.nih.gov/articles/PMC9500287/pmc.ncbi.nlm.nih+1

Chapter 23

254 "Voluntary Service Overseas (VSO)." *Voluntary Service Overseas – Wikipedia* (accessed 2026).
Web: https://en.wikipedia.org/wiki/Voluntary_Service_Overseas

255 "Social Change Initiative." *The Social Change Initiative – What We Do* (accessed 2026).
Web: https://www.socialchangeinitiative.com/what-we-do

256 Othering & Belonging Institute. "Democracy & Belonging Forum." University of California, Berkeley. Accessed March 15, 2026. https://belonging.berkeley.edu/democracy-belonging-forum.belonging.berkeley

257 United Nations Development Programme. "About Us." Accessed March 15, 2026. https://www.undp.org/about-us.

258 Council of Europe "No Hate Speech Movement" overview: https://www.europewatchdog.info/en/instruments/campaigns/no-hate-speech/europewatchdog

259 Australian Human Rights Commission. "Racism. It Stops With Me." Accessed March 15, 2026. https://humanrights.gov.au/know-your-rights/rights-of-individuals/race-discrimination/racism-it-stops-me-campaign-relaunch.humanrights

260 Elon Musk, "The reality is great highs, terrible lows and unrelenting stress. Don't think people want to hear about the last two," X (formerly Twitter), July 30, 2017, 10:37 a.m., https://x.com/elonmusk/status/891710778205626368.avc+2

261 Russ Mitchell, "Is Elon Musk a Bad Boss? Ask Tesla, SpaceX, Twitter Workers," Los Angeles Times, November 13, 2022, https://www.latimes.com/business/story/2022-11-14/elon-musk-toxic-boss-timeline.

262 Grace Kay, "Elon Musk Ruthlessly Fired Anyone Who Disagreed With Him, New Book Says," Business Insider, August 2, 2021, https://www.businessinsider.com/tesla-elon-musk-ruthlessly-fired-anyone-who-disagreed-spacex-report-2021-8.businessinsider

263 Michael C. Bender and Kenneth P. Vogel, "Elon Musk Backed Trump With Over $250 Million, Fueling the 'RBG PAC'," New York Times, December 5, 2024, https://www.nytimes.com/2024/12/05/us/politics/elon-musk-trump-rbg-election.html.nytimes

264 Robert Hart, "Elon Musk Says Neuralink Could Slash Risk From AI as Firm Prepares for First Human Trials," Forbes, September 21, 2023, https://www.forbes.com/sites/roberthart/2023/09/21/elon-musk-says-neuralink-could-slash-risk-from-ai-as-firm-prepares-for-first-human-trials/.forbes

265 Mark Kamine, "Elon Musk Says Neuralink Has Implanted a Brain Chip in Second Patient," Mashable, August 5, 2024, https://mashable.com/article/neuralink-second-patient-elon-musk.mashable

266 Rachael Levy and Marisa Taylor, "Musk's Neuralink Faces Federal Probe, Employee Backlash over Animal Tests," Reuters, December 5, 2022, https://www.reuters.com/technology/musks-neuralink-faces-federal-probe-employee-backlash-over-animal-tests-2022-12-05/.reuters

267 FDA Purge Included Workers Reviewing Elon Musk's Neuralink Brain Implant Company," Democracy Now!, February 18, 2025, https://www.democracynow.org/2025/2/18/headlines/fda_p

urge_included_workers_reviewing_elon_musks_neuralink_brain_implant_company.democracynow

268 Kwan Wei Kevin Tan, "Elon Musk Blocked Ukraine's Starlink Access near Crimea, Thwarting a Major Offensive against Russia, Book Says," Business Insider, September 6, 2023, https://www.businessinsider.com/elon-musk-blocked-ukraine-starlink-access-crimea-russia-war-putin-2023-9.businessinsider

269 Timothy Graham and Mark Andrejevic, "Elon Musk Appears to Have Tweaked X's Algorithm to Boost His Posts and Pro-Trump Content," Independent, November 3, 2024.independent.co+1

270 Ian Sherr, "Musk Reportedly Pushed Algorithm Change to Boost His Tweets," CNET, February 14, 2023, https://www.cnet.com/news/social-media/musk-reportedly-pushed-algorithm-change-to-boost-his-tweets/.cnet

Chapter 25

271 Loredana Fattorini et al., "The Global AI Vibrancy Tool," Stanford Institute for Human-Centered Artificial Intelligence (HAI), November 2024, https://hai.stanford.edu/ai-index/global-vibrancy-tool.hai.stanford+1

272 Raenette Taljaard Adams et al., *Global Index on Responsible AI 2024 (1st ed.)* (Johannesburg: Global Center on AI Governance, 2024), https://www.globalcenter.ai/research/2024-global-index-on-responsible-ai.scribd+1

273 "AI Act Enters into Force," European Commission, July 31, 2024, https://commission.europa.eu/news-and-media/news/ai-act-enters-force-2024-08-01_en.commission.europa

274 OECD, "AI Principles," accessed March 16, 2026, https://www.oecd.org/en/topics/ai-principles.html.

275 Global Partnership on Artificial Intelligence," OECD, accessed March 16, 2026, https://www.oecd.org/en/about/programmes/global-partnership-on-artificial-intelligence.html.

276 "About the Pact," *Pact for the Future – The United Nations*, accessed March 16, 2026, https://www.un.org/pact-for-the-future/en/about-pact.idea+1

• 277 Joseph R. Biden Jr., "Executive Order 14110: Safe, Secure, and Trustworthy Development and Use of Artificial Intelligence," October 30, 2023, in *Federal Register* 88, no. 209 (November 1, 2023).bidenwhitehouse.archives+2. https://www.reginfo.gov/public/do/DownloadDocument?objectID=139156001reginfo

278 Nita A. Farahany, *The Battle for Your Brain: Defending the Right to Think Freely in the Age of Neurotechnology* (New York: St. Martin's Press, 2023).magazine.law.duke+2. https://www.nitafarahany.com/the-battle-for-your-brainnitafarahany

279 Rafael Yuste et al., "Neurorights in the Constitution: From Neurotechnology to Ethics and Politics," *Philosophical Transactions of the Royal Society B* 379, no. 1907 (2024): 20230098.royalsocietypublishing+2. https://royalsocietypublishing.org/doi/10.1098/rstb.2023.0098royalsocietypublishing

280 Global Neuroethics Summit Delegates, "Neuroethics Questions to Guide Ethical Research in the International Brain Initiatives," Neuron 100, no. 1 (2018): 19–36, https://www.sciencedirect.com/science/article/pii/S0896627318308237.sciencedirect+1

281 Sawai, Tetsu, et al. "The Ethics of Human Brain Organoid Transplantation in Animals." *Cell Stem Cell* 30, no. 1 (2023): 10–24. https://pmc.ncbi.nlm.nih.gov/articles/PMC10550858/.

282 Susan Schneider, "Merging with AI Would Be Suicide for the Human Mind," *Financial Times*, August 12, 2019, https://www.ft.com/content/0c4fac58-bd15-11e9-9381-78bab8a70848.ft

283 E. C. Gordon et al., "Ethical Considerations for the Use of Brain–Computer Interfaces for Cognitive Enhancement," *PLOS Biology* 22, no. 10 (October 27, 2024): e3002899, https://journals.plos.org/plosbiology/article?id=10.1371/journal.pbio.3002899.

284 Pycroft et al., "Brainjacking: Implant Security Issues in Invasive Neuromodulation," *World Neurosurgery* 92 (2016): 454–

462, https://www.sciencedirect.com/science/article/abs/pii/S1878
875016302728.sciencedirect

Index

About the Author

Dr. John Sydney Smith, MD, is a neuropsychiatrist and medico-legal consultant recognized for his work in the evaluation of cognitive and behavioral disorders. He has served as the inaugural Director of the Neuropsychiatric Institute at the University of New South Wales and contributed to advancing the field of neuropsychiatry in both clinical and legal contexts. Dr. Smith's expertise includes complex assessments within forensic settings, highlighting his reputation as a leading authority in medico-legal psychiatry.

Throughout his distinguished career, Dr. Smith has engaged in both research and practice, co-authoring publications on dementia, brain damage and surgical interventions for epilepsy and severe psychiatric disorders and sharing insights into the interplay between neurological conditions and behavior. He is noted for his commitment to human rights and ethical practice within psychiatry, having participated in significant professional and public discussions on patient care standards.

Dr. Smith's broad influence within the neuropsychiatric and psychiatric professions stems from his leadership roles, scholarly contributions, and dedication to the humane and scientifically grounded advancement of mental health care.

www.ingramcontent.com/pod-product-compliance
Lightning Source LLC
Chambersburg PA
CBHW070730030726
47601CB00002B/216